ELECTRONIC TECHTONICS:
THINKING AT THE INTERFACE

Proceedings of the
First International HASTAC Conference
Duke University, Durham, North Carolina
April 19-21, 2007

EDITORS:

Erin Ennis
Zoë Marie Jones
Paolo Mangiafico
Mark Olson
Jennifer Rhee
Mitali Routh
Jonathan E. Tarr
Brett Walters

This conference was organized by the "Interface" Seminar of the
John Hope Franklin Humanities Institute at Duke University as its
contribution to HASTAC's In|Formation Year.

Published by Lulu Press
http://stores.lulu.com/hastac

Cover design by Jason Doty.

ISBN 978-1-4357-1362-8

CONTENTS

Panel 4: Electronic Book Review 4.0: Toward a Semantic Literary Web-Based Interface

Panel 5: The World Wide Web Evolves**

Panel 6: Racing (through) Domains

Panel 7: Connecting the (Virtual) Dots

Panel 8: Innerspace and Interface

Panel 9: Ludic Depths: Games, Narratives, Platforms

* The audio for this panel is available at www.hastac.org.
** Available in digital format in the video archives of the conference at
www.hastac.org.

Preface

This volume originated in HASTAC's first international conference, "Electronic Techtonics: Thinking at the Interface," held at Duke University during April 19-21, 2007. "Electronic Techtonics" was the site of truly unforgettable conversations and encounters that traversed domains, disciplines, and media – conversations that explored the fluidity of technology both *as* interface as well as *at* the interface.

This hardcopy version of the conference proceedings is published in conjunction with its electronic counterpart (found at www.hastac.org). Both versions exist as records of the range and depth of conversations that took place at the conference. Some of the papers in this volume are almost exact records of talks given at the conference, while others are versions that were revised and reworked some time after the conference. These papers are drawn from a variety of fields and we have not made an effort to homogenize them in any way, but have instead retained the individual format and style of each author.

This hardcopy volume does not contain every work presented at the conference. While the papers included in this volume can be found in both hardcopy and electronic publications, versions of other events and presentations can only be found in the electronic volume of the proceedings. For example, the four keynote speeches (by John Seely Brown, James Boyle, John Unsworth, and Rebecca Allen) are not included in this volume. Likewise, some of the events listed in the conference program/table of contents exist in media forms more amenable to electronic publication, such as the panel discussions. Both the keynote addresses and the panel discussions are available in digital format in the video archives of the conference (www.hastac.org).

In publishing the proceedings of the HASTAC conference as such (collaboratively across media), we hope to capture some of the innovation, dynamism, and creative urgency that invigorated the three days of the conference itself.

-- The Editorial Collective

ACKNOWLEDGMENTS

HASTAC would like to acknowledge the generous support of our sponsors and the extraordinary effort of those who made possible our first international conference.

Sponsors:
Duke University
> Office of the President, Office of the Provost, Vice Provost for Academic Affairs, Office of Information Technology, John Hope Franklin Humanities Institute, John Hope Franklin Center for Interdisciplinary and International Studies, Program in Information Science + Information Studies (ISIS), Nasher Museum of Art, Department of Art, Art History, and Visual Studies, School of Nursing, Fitzpatrick Center for Interdisciplinary Engineering, Medicine, and Applied Sciences (FCIEMAS).

John D. and Catherine T. MacArthur Foundation
> MacArthur's $50 million digital media and learning initiative seeks to gain a better understanding of how digital technologies are changing how young people learn, play, socialize, exercise judgment, and engage in civic life.

Renaissance Computing Institute (RENCI)
The Andrew W. Mellon Foundation Sawyer Seminar on Human Being, Human Diversity, and Human Welfare
The National Science Foundation
Digital Promise

Conference Committee:
Cathy N. Davidson, Chair
Jonathan E. Tarr: HASTAC Project Manager
Nihad Farooq: Conference Session and Panel Coordinator
Erin Ennis, HASTAC Conference Special Assistant
Pamela Gutlon: Site Consultant
Mark Olson: Director of New Media and Information Technology
Brett Walters: HASTAC Webmaster
Jason Doty: Graphic Designer
Harry Halpin: Demo/Exhibits Support

Program Selection Committee: Interface Seminar, John Hope Franklin Humanities Institute, Duke University. Co-conveners, Tim Lenoir (Kimberly Jenkins Chair in New Technologies and Society) and Priscilla Wald (English).

Members: Anne Allison (Cultural Anthropology), Rachael Brady (Visualization Technology Group, Electrical and Computer Engineering),

Cathy N. Davidson (Franklin Humanities Institute Professor of Interdisciplinary Studies), Guven Guzeldere (Philosophy, Linguistics, Neurobiology, Psychology, and Neuroscience), Orit Halpern (Postdoctoral Fellow, Historical Studies), Andrew Janiak (Philosophy), Harry Halpin (Philosophy and Computer Science), David Liu (Religion), Marilyn Lombardi (Office of Information Technology), Paolo Mangiafico (Digital Projects, Perkins Library), Robert Mitchell (English), Mark Olson (New Media and Information Technology, John Hope Franklin Center), Jennifer Rhee (Literature), Mitali Routh (Art, Art History and Visual Studies), and Kristine Stiles (Art, Art History and Visual Studies).

Keynote Speakers: John Seely Brown, University of Southern California; James Boyle, Duke University; John Unsworth, University of Illinois, Urbana-Champaign; and Rebecca Allen, University of California, Los Angeles.

A special thank you for support and participation to the John D. and Catherine T. MacArthur Foundation: Julia M Stasch, Vice President, Human and Community Development; Constance M. Yowell, Director of Education.

Thank you to Durham Public Schools: Superintendent Carl Harris; Nancy Hester, Associate Superintendent; Terri Mozingo, Chief Academic Officer.

Special thanks also to the following programs and individuals at Duke University: the Franklin Humanities Institute: Grant Samuelsen, Associate Director; Christina Chia, Assistant Director for Communications; and Robin Geller, Program Coordinator; the John Hope Franklin Center for Interdisciplinary and International Studies; the Office of Public Affairs and Government Relations: John Burness, Senior Vice President for Public Affairs and Government Relations; the Office of News & Communications: Sally Hicks, Senior Writer; President Richard H. Brodhead; Richard Riddell, Vice President and University Secretary; Peter Lange, Provost; John Simon, Vice Provost for Academic Affairs; George L. McLendon, Dean of Arts and Sciences; N. Gregson G. Davis, Dean of the Humanities; Kristina Johnson, former Dean of the Pratt School of Engineering and current Provost of Johns Hopkins University; Sarah Deutsch, Dean of Social Sciences; Robert J. Thompson, Jr., Dean of Trinity College; Stephen Nowicki, Dean of Undergraduate Education; Kimerly Rorschach, Mary D.B.T. and James H. Semans Director of the Nasher Museum of Art; Catherine Lynch Gilliss, Dean of the School of Nursing.

* * *

"Electronic Techtonics: Thinking at the Interface" is part of the HASTAC In | Formation Year, a collaboration by over eighty centers, institutes, universities, libraries, and museums. The In | Formation Year is exploring the humane and humanistic dimensions of technology and introducing an array of new technological innovations. Please see www.hastac.org for a list of all people and institutions who made this year possible as well as for free, downloadable archives of the year's broadcasts. We offer special thanks to the following site leaders and institutions: Anne Balsamo (USC), Ruzena Bajcsy (UC Berkeley), Allison Clark (UIUC), Cathy N. Davidson (Duke), David Theo Goldberg (UCHRI), Daniel Herwitz (Michigan), Julie Thompson Klein (Wayne State), Henry Lowood (Stanford), Thomas MacCalla (National), Tara McPherson (USC), Katherine Mezure (Mills College), and Kathleen Woodward (Washington).

ABOUT HASTAC

HASTAC (pronounced "haystack") is an acronym for Humanities, Arts, Science, and Technology Advanced Collaboratory. It is an entirely voluntary consortium of leading researchers and nonprofit research institutions worldwide. Its primary members are universities, supercomputing centers, grid and teragrid associations, humanities institutes, museums, libraries, and other civic institutions.

HASTAC was founded by Cathy N. Davidson, former Vice Provost for Interdisciplinary Studies and co-founder of the John Hope Franklin Humanities Institute at Duke University, and David Theo Goldberg, Director of the University of California's state-wide Humanities Research Institute (UCHRI). At a meeting of humanities leaders held by the Mellon Foundation in 2002, it was clear that Davidson and Goldberg had been working on a variety of projects with leading scientists and engineers dedicated to expanding the innovative uses of technology and to thinking together about social, ethical, and access issues of cyberinfrastructure in parallel with the process of creating it. Each of them also knew of leaders at other institutions who shared that vision and, within a few months, the HASTAC consortium was born.

The HASTAC network consists of more than eighty institutions principally located in the US and reaches over 30,000 people worldwide. In reality, it is more a network of networks, located at the intersection of technology, engineering, and computing on one hand, and the humanities, arts and social sciences on the other. This profound interconnectivity has allowed HASTAC to develop its successful network, which in turn promotes greater interactive connections.

HASTAC 2006-07 Steering Committee
Ruzena Bajcsy, Director Emerita of the Center for Information Technology Research in the Interest of Society (CITRIS) at the University of California, Berkeley
Anne Balsamo, Professor, Interactive Media and Gender Studies and Managing Director, Institute for Multimedia Literacy, University of Southern California
Allison Clark, Associate Director, Seedbed Initiative for Transdomain Creativity, University of Illinois, Urbana-Champaign
Cathy N. Davidson, Professor of Interdisciplinary Studies, John Hope Franklin Humanities Institute, and Ruth F. DeVarney Professor of English at Duke University
Kevin Franklin, Executive Director, Center for Computing in the Humanities, Arts, and Social Science (CHASS), University of Illinois, Urbana-Champaign

David Theo Goldberg, Director of the University of California Humanities Research Institute (UCHRI) and Professor of African-American Studies and of Criminology, Law, and Society at the University of California, Irvine

Daniel Herwitz, Director and Mary Fair Croushore Professor of Humanities, University of Michigan

Julie Thompson Klein, Professor of Humanities in Interdisciplinary Studies, and Faculty Fellow in the Office of Teaching & Learning and Co-Director of the University Library Digital Media Project, Wayne State University

Henry Lowood, Curator for Germanic Collections and History of Science and Technology Collections at Stanford University Libraries, Stanford University

Thomas Maccalla, Executive Director, National University Community Research Institute (NUCRI) and University Vice President, National University

Stephenie McLean, Director of Education and Outreach, Renaissance Computing Institute (RENCI)

Tara McPherson, Editor, Vectors, and Associate Professor of Critical Studies in the School of Cinema-Television at the University of Southern California

Mark Olson, Director, New Media and Information Technologies, John Hope Franklin Center for Interdisciplinary and International Studies, Duke University

Douglas Thomas, Associate Professor, Annenberg School for Communication, University of Southern California

Kathleen Woodward, Professor of English, Director of the Walter Chapin Simpson Center for the Humanities, University of Washington

The 1st International HASTAC Conference
April 19-21, 2007*

Thursday, April 19, 2007, 8:00 p.m., Keynote Address: John Seely Brown, "The Social Life of Learning in the Net Age," Nasher Museum Auditorium

Introduction: Timothy Lenoir, Jenkins Chair in New Technologies and Society, and Co-Convener, "Interface" Seminar, Franklin Humanities Institute, Duke University.

John Seely Brown is one of the formative thinkers of the Information Age. He is currently a visiting scholar at the University of Southern California. Prior to that he was the Chief Scientist of Xerox Corporation and the director of its Palo Alto Research Center (PARC), a position he held for two decades. While head of PARC, Brown espoused radical innovation, expanding the role of corporate research to include such topics as organizational learning, knowledge management, complex adaptive systems, ethnographic studies of the workscape, and nanotechnology. He was also a co-founder of the Institute for Research on Learning (IRL). His personal research interests include the impact of globalization on business, the management of radical innovation, digital culture, ubiquitous computing, and organizational and individual learning. He is the author of several books and over a hundred scientific papers and, with Paul Duguid, wrote the transformative work, *The Social Life of Information* (2000).

9:00 p.m. - Reception in Nasher Museum Atrium. Music by Steve Burnett, thereminist. Exhibit pavilions will be open during the reception.

Friday, April 20, 2007 > Talks, Panels, Exhibits, Demos > Marriott Hotel and Durham Arts Council

8:00-9:00 a.m. - Continental Breakfast, First Floor Hall

9:00-10:00 a.m. - James Boyle: **Creative Commons, Science Commons, and Open Source** - Ballroom 103.

Introduction: Priscilla Wald, Department of English and Co-Convener, "Interface" Seminar, Franklin Humanities Institute, Duke University James Boyle is William Neal Reynolds Professor of Law and co-founder of the Center for the Study of the Public Domain at Duke Law School. He is one of the founding Board Members of Creative Commons, which is working to facilitate the free availability of art, scholarship, and cultural materials by developing innovative, machine-readable licenses that individuals and institutions can attach to their work, and of Science

---

* This program reflects the original schedule for the 1st International HASTAC Conference. The titles of some papers have changed for this publication.

Commons, which aims to expand the Creative Commons mission into the realm of scientific and technical data. Professor Boyle is the author of numerous works, including *Shamans, Software and Spleens: Law and Construction of the Information Society* (Harvard University Press 1996) and, most recently, co-author of *Bound By Law* (CSPD 2006), a comic book on fair use in documentary film. He is the winner of the 2003 World Technology Award for Law for his work on the "intellectual ecology" of the public domain, and on the new "enclosure movement" that seems to threaten it. He currently writes as an online columnist for the Financial Times' New Economy Policy Forum.

10:15-11:30 a.m. Breakout sessions:

#1. Funding the Digital Future: *Leaders from national agencies, private foundations, and industry discuss digital funding opportunities, initiatives, and visions.* Meeting Room 108. Session Chair: Julie Thompson Klein, Wayne State University.
Brett Bobley, CIO and Director of Digital Humanities Initiative, National Endowment for the Humanities
Karl Brown, Associate Director, Applied Technology, The Rockefeller Foundation
Jerry Heneghan, CEO, Virtual Heroes, and Chairman, North Carolina Association for Advanced Learning Technologies (NCALTA)
Gary Kebbel, Journalism Program Officer, John S. and James L. Knight Foundation
Kevin M. Guthrie, President, Ithaka; Diana Rhoten, Program Director, Office of Cyberinfrastructure, National Science Foundation
Steven C. Wheatley, Vice President, American Council of Learned Societies
Constance M. Yowell, Director for Digital Media, Learning, and Education, John D. and Catherine T. MacArthur Foundation

#2. Interface Genealogies: *Previous moments in media history illuminate what is and isn't new about "New Media."*
Meeting Room 105. Session Chair: Jennifer Rhee, Program in Literature, Duke University. Discussion Leader: Andrew Janiak, Department of Philosophy, Duke University.
Caitlin Fisher, Department of Film, York University, Canada, "Interface Epistemology: Hypermedia Work in the Academy"
Lisa Gitelman, Department of Media Studies, Catholic University, "Xerographers of the Mind: The Lost Idea of the Photocopy"
Matthew Tinkcom, Graduate Program in Communication, Culture, and Technology, Georgetown University, "Eduction: A Theory of Value in the Digital Cinematic Epoch"

#3. Theorizing Interface: *Metaphors help us comprehend how digitality weaves, binds, encloses, bridges, spans, and navigates across technologies, spaces, and disciplines (from genomics to urban planning).*
Meeting Room #106. Session Chair: David Liu, Department of Religion, Duke University. Discussion Leader: Lev Manovich, Department of Visual Arts, University of California, San Diego.
Sylvia Nagl & Sally Jane Norman, Department of Oncology, University College London, Culture Lab, Newcastle University, "Raranga Tangata: The Weaving Together of People"
Nicole Starosielski, Department of Film and Media Studies, University of California, Santa Barbara, "Reskinning the Digital Surface: Borders and Immobility at the Interface"
Sarah Sweeney, Digital Media Arts Program, Mercer County Community College, "Way-Finding on the Web: Urban Planning and the Virtual Interface"

#4. Electronic Book Review 4.0: Toward a Semantic Literary Web-based Interface: *The Electronic Book Review showcases experiments in design, intellectual property, authorship, semantics, taxonomy, and reading practices.*
Meeting Room #107. Session Chair: Robert Mitchell, Department of English, Duke University. Discussion Leader: Victoria Szabo, Program in Information Science + Information Studies, Duke University.
Joseph Tabbi, Editor, EBR, "Toward a Semantic Literary Web"
Ewan Branda, Database and Application Designer, EBR, "A Map of Relations: the Software and Data Architecture of EBR 4.0"
Anne Burdick, Interface Designer, EBR, "EBR 4.0: The Interface as a Tool for Reading & Writing"

11:00 AM-5:00 p.m. **Interfaces of the Future**: Exhibits and demos. Ballroom 103. Continuous and scheduled demos. Details in conference packet.

12:00-2:00 p.m. - Lunch and Panel: **The Foundations and Futures of Digital Humanities**: Discussion led by John Unsworth. Ballroom 101/102.
Introduction: Kathleen Woodward, Department of English and Director, Simpson Center for the Humanities, University of Washington
John Unsworth is Dean of the Graduate School of Library and Information Science (GSLIS) at the University of Illinois, Urbana-Champaign. Previously, he served as the Director of the Institute for Advanced Technology in the Humanities at the University of Virginia. For his work at IATH, he received the 2005 Richard W. Lyman Award from the National Humanities Center. He co-chaired the national commission that produced the 2006 report on Cyberinfrastructure for Humanities and Social Science, on behalf of the American Council of Learned Societies, and he has supervised research projects across the disciplines in the

humanities.

Panelists:

> Susan Brown, Orlando Project, University of Guelph
> Kathleen Fitzpatrick, MediaCommons, Pomona College
> Henry Lowood, Stanford Humanities Lab, Stanford University
> Tara McPherson, Vectors, University of Southern California
> Catherine Mitchell, California Digital Library, University of California Libraries
> Kenneth M. Price, Center for Digital Research in the Humanities and The Whitman Archive, University of Nebraska-Lincoln
> Martha Nell Smith, Maryland Institute for Technology in the Humanities, University of Maryland

2:30-4:30 p.m. Breakout sessions:

#5. The World Wide Web Evolves *Formative figures in the creation of the current Web--semantic Web, the grid, and social software--envision Web 3.0.* Meeting Room 106. Session Chair: Paolo Mangiafico, Digital Projects, Duke University Libraries. Discussion Leader: Harry Halpin, Electrical and Computer Engineering Department, Duke University, and School of Informatics, University of Edinburgh.

Dan Connolly, Research Scientist, MIT Computer Science and Artificial Intelligence Laboratory and World Wide Web Consortium (W3C), "How the W3C Process Got Its Stripes"

Pat Hayes and **Margaret Warren**, Senior Research Scientist, Institute for Human and Machine Cognition, Founder, CARMA (Cyber Arts, Research Music and Audio Productions), "Artspeak: The Contemporary Artist meets the Semantic Web. Creating Formal Semantic Web Ontologies from the Language of Artists"

David de Roure, Head of Grid and Pervasive Computing in the School of Electronics and Computer Science, University of Southampton, "Grid of People"

Henry Thompson, Human Communication Research Centre, University of Edinburgh and World Wide Web Consortium (W3C), "The Humanities, the New Empiricism, and the World Wide Web"

#6. Racing (through) Domains *Racial attitudes persist in digital media and in race-based surveillance but also in new methods for teaching civil rights history.* Meeting Room 107. Session Chair: Anna Everett, Department of Film Studies, University of California - Santa Barbara. Discussion Leader: Allison Clark, Seedbed Initiative for Transdomain Creativity, University of Illinois, Urbana-Champaign.

Jessie Daniels, Urban Public Health Program/Community Health Education, CUNY, Hunter College, "Cloaked Websites, Youth, and Digital Media: Thinking about Race and Civil Rights at the Interface"

Simone Browne, Department of Sociology and Equity Studies in

Education, University of Toronto, "(Im)mobility Documents, Race, and Surveillance"

Irene Chien, Film Studies and New Media Program, University of California, Berkeley, "Orienting Inner Space: Biofeedback Gaming and the Racialized Landscape of Mind, Body, and Spirit"

Michele White, Department of Communications, Tulane University, "The Hand Blocks the Screen: A Consideration of the Ways the Interface is Raced"

#7. Connecting the (Virtual) Dots *Simulations, emergence, augmented life, and visualization technologies animate cultural spaces, historical enterprises, games, and corpora as well as the military.*
Meeting Room 105. Session Chair: Orit Halpern, Franklin Humanities Institute, Duke University, and Department of Historical Studies, New School for Social Research. Discussion Leader: Mitali Routh, Department of Art, Art History, and Visual Studies, Duke University.

Timothy R. Tangherlini, **Zoe Borovsky**, & **Todd Presner**, UCLA Center for Digital Humanities, UCLA Digital Humanities Incubation Group, "Thick Viewing: Integrated Visualization Environments for Humanities Research on Complex Corpora"

Helen Papagiannis, Joint Program in Communication and Culture, York University and Ryerson University, "Augmenting Digital and Analog Memory"

John H. Johnston, Department of English, Emory University, "Artificial Life: New Media Object as a New Space of Exploration"

Caren Kaplan, Cultural Studies Program (Women and Gender Studies), University of California, Davis, "'Everything is Connected': Aerial Perspectives, the 'Revolution in Military Affairs,' and Digital Culture"

#8. Innerspace and Interface *Affect and representation are crucial to digital history, music, and dance.*
Meeting Room 108. Session Chair: Marilyn Lombardi, Office of Information Technology, Duke University. Discussion Leader: Thomas MacCalla, Community Research Institute, National University.

Jennifer Boyle, Department of English, Hollins University and Carol G. Lederer, Fellow, Pembroke Center, Brown University, "The Hollins Community Project: Interfacing Affect"

John Toenjes & **David Marchant**, Department of Dance, University of Illinois, Urbana-Champaign, Performing Arts Department, Washington University-St. Louis, "Finding Humanity Within the Machine: Large Motor Movement Computer Interfacing as an Artistic Mindbody Integrative Practice"

Ulrich Rauch & **Tim Wang**, Arts Instructional Support and Information Technology Group, University of British Columbia, Instructional Development, University of British Columbia, "Art Spaces: Reconstructing the Past"

4:30 p.m. Open time, Downtown Durham. *Demos will be on open for screening. Conference participants are invited to visit downtown galleries, artist spaces, pubs, and cafés (maps provided).*

6:00-8:00 p.m. - Cash bar at Marriott

6:30-7:30 p.m. - Conference Banquet - Ballroom 101/102

7:30-8:30 p.m. - Rebecca Allen: **Global Interfaces, Intimate Interfaces and the Interface between Art and Technology**
Introduction: David Theo Goldberg, Director, University of California Humanities Research Institute and HASTAC Co-Founder

9:00-11:00 p.m. – René Garcia (VJ Cyops) - **ReMix2** with DJ RasMusis. PSI Theatre, Durham Arts Council. *Performance artist and VJ Rene Garcia (VJ Cyops) will create a live remix of video and soundscapes to propel us through critiques of race, terrorism, and the neo-surveillance state to a dance-hall evening of community, activism, and resistance.*

Saturday, April 21. Talks, Panels, Exhibits, Demos. Duke University: School of Nursing (307 Trent Drive), John Hope Franklin Center (corner of Erwin Rd. and Trent Dr.), FCIEMAS (100 and 101 Science Dr.).
The morning events are open to conference registrants and to the public (free of charge and on a space available basis).

8:30-9:00 a.m. - Continental Breakfast, Atrium & Patio, Duke University School of Nursing Building

9:00-10:30 a.m. - **The Future of Learning: Three Perspectives**
Introductory Remarks: Provost Peter Lange, Duke University (DukeEngage)
Panelists:
> Julia Stasch, Vice President, Human and Community Development, John D. and Catherine T. MacArthur Foundation, "Building the Field of Digital Media and Learning" (www.digitallearning.macfound.org)
> Cathy N. Davidson and David Theo Goldberg, "The Future of Learning Institutions in a Digital Age" (www.futureofthebook.org/HASTAC/learningreport/about/url)
> Dr. Carl Harris, Superintendent, Durham Public Schools, "A Public School Perspective on the Future of Learning"

Discussion Leader:
> Connie Yowell, Director for Digital Media, Learning and Education, MacArthur Foundation

10:45 a.m.-12:30 p.m. **At the Interface of Everything** *A rare conversation across domains among digital visionaries. The outcome will be a mind-map of the conference and a game-plan for unforseeable futures.*

Moderator:

> Anne Balsamo, "Technohumanist," Institute for Multimedia Literacy and the Collaboratory for Technology and Culture, University of Southern California

Participants:

> Rebecca Allen, UCLA, *new media design, universal access*
> Ruzena Bajcsy, UC-Berkeley, *tele-immersive environments*
> James Boyle, Duke University Law School, *creative commons, science commons, open source*
> Rachael Brady, Duke University, *scientific visualization*
> John Seely Brown, Former Xerox Chief Scientist and Director, Xerox PARC, *radical innovation*
> Jonathon Cummings, Fuqua School of Business, Duke University, *distributed research teams, collaboration*
> Dan Connolly, MIT, *W3C technical architecture*
> Anna Everett, UC-Santa Barbara, *media and race theory*
> Kevin Franklin, UC Humanities Research Institute, *global access grid*
> Lev Manovich, UC-San Diego, *new media art and theory*
> Fred Stutzman, UNC-Chapel Hill, *social networks research*
> Douglas Thomas, USC, *cultural studies and conceptual blending*

The below afternoon events are open to conference registrants only

12:30-2:00 p.m. Informal buffet lunch. 240 Franklin Center. Lunchtime conversation, **The Future of Art in a Digital Age** *Visual, sound, and multimedia artists (whose work will be performed or shown throughout the conference) address the problems and potentials of making art in a technological age.*

Session Chair:

> Kristine Stiles, Department of Art, Art History, and Visual Studies, Duke University.

Participants:

> Anya Belkina, Visual Artist, Rumi, Duke University
> J-Bully (a.k.a. Robi Roberts), Rapper, MiX TAPEStry, Duke University
> René Garcia, VJ, Video artist University of Southern California, Los Angeles
> Suguru Goto, Bodysuit, Robotic Music, Visiting Artist, Ohio University
> Scott Lindroth, Composer, Rumi, MiX TAPEstry, Duke University
> Mendi + Keith Obadike, Music, Live Art, Conceptual Internet Art, Princeton University and William Paterson University

2:00-3:15 p.m.

#9. Ludic Depths: Games, Narratives, Platforms *Complex and sometimes contradictory notions of narrative play out in hardware and software design, game structures, and historical modeling and pedagogy.*
Room 1011, School of Nursing. Session Chair: Victoria Szabo, Program in Information Science + Information Studies, Duke University. Discussion Leader: Patrick Jagoda, Department of English, Duke University.
Ian Bogost & **Nick Montfort**, School of Literature, Communication and Culture, Georgia Institute of Technology, Department of Computer and Information Science, University of Pennsylvania, "New Media as Material Constraint: An Introduction to Platform Studies"
Patricia Seed, Department of History, University of California, Irvine, "Learning History by Designing Games: A New Approach to Teaching History"
Noah Wardrip-Fruin, Department of Communications, University of California, San Diego, "Internal Processes and Interface Effects: Three Relationships in Play"

2:00-5:00 p.m. Guided Tours for General Attendees (groups of 10-20). Demos include presentations by the artists or developers, screenings, and interactive experiences. John Hope Franklin Center:
-   Franklin Center Media Gallery: "Ex Machina" installation by artists Christian Karkow: interactive sculpture.
-   Franklin Center Main Gallery, John Hope Franklin Center: "On Reading" Exhibit by Wendy Ewald.
-   Screening of performance of *Nasuh* bu Rumi, multimedia concert by artist Anya Belkina and composer Scott Lindroth.
-   Screening of internet and multimedia art work by Mendi + Keith Obadike.
-   Screening of performances by Suguru Goto of *BodyWorks* and *Robot Music*.
-   Games Exhibit in the Interactive Multimedia Project Space (IMPS). Student-designed interactive games, Virtual Nasher, and demo of pedagogical uses from the FOCUS *Games2Know* cluster and How *They Got Game*. IMPS and applications designed by Mark Olson, Tim Lenoir, and Zach Pogue.
-   Video of MiX TAPEStry by J-Bully (which premiered at InCommon, *Katrina: After the Storm*). Inspired by Allison Clark, music by Scott Lindroth, and graphics by John Jennings.
-   Fitzpatrick Center for Interdisciplinary Engineering, Medicine, and Applied Sciences (FCIEMAS): "Future and History of the Interface": Designed specifically for the Interface conference, "Future and History of the Interface" encompasses two whole-body interfaces to the same conceptual dataset. A fully immersive virtual reality experience is offered in the DiVE tank. At this location, visually

complex patent and citation networks are presented in two parallel planes. The second whole-body interface is offered in The Studio using motion detection. Three narratives depicting independent pathways leading out of the work from Xerox PARC are triggered through the clustering and collaboration of individuals within The Studio.

- Demonstration of OpenCroquet, a next-generation powerful open source software development environment for the creation of and large-scale distributed deployment of multi-user virtual 3D applications and metaverses. The OpenCroquet Project was founded by Alan Kay, David Smith, David Reed, Julian Lombardi, Mark McCahill, and Andreas Raab.

Opening Remarks

Cathy Davidson

WELCOME! I am Cathy Davidson, Interim Director of the John Hope Franklin Humanities Institute at Duke University and a co-founder of HASTAC, along with David Theo Goldberg, Director of the University of California's Humanities Research Institute. H-A-S-T-A-C is an acronym that stands for "Humanities, Arts, Sciences, and Technology Advanced Collaboratory." Everyone just says "haystack." It is my pleasure and privilege to be able to welcome you tonight.

Welcome to our talk this evening by John Seely Brown, the intellectual and spiritual godfather of the Information Age and, we are pleased to say, of HASTAC. His talk will be followed by a reception in the Nasher atrium, with live music by thereminist Steve Burnett, to which you are all invited. This is the first of three events open to the public as "The Future of Learning." The orange programs will provide you with information about the other events to follow on Saturday.

Welcome to the opening of "Electronic Techtonics: Thinking at the Interface," our first-ever international HASTAC conference. The three "Future of Learning Events" are the public events sponsored by HASTAC as part of our first international gathering. Tomorrow marks the beginning of our events for conference registrants. If you are a conferee and have not received a copy of the summary program, you can pick one up at the reception tonight.

Welcome to those watching this webcast to the eighth of our nine distributed "In|Formation Year" events sponsored by HASTAC and over eighty affiliated universities, humanities centers, science centers, museums, libraries, and civic organizations who have been responsible for a full-year of programming around In|Formation themes. The themes and host institutions are: InCommon (led by UIUC), Interplay (USC), In Community (National University), Interaction (UC Berkeley, Mills College, and Stanford), Integration (Wayne State), Injustice (Michigan), Invitation (Washington), Interface (Duke), and, coming on May 10, Innovation (sponsored by UCHRI). We believe all of those topics—from injustice to innovation—are key to understanding our Information Age.

Welcome to the culminating event of our John Hope Franklin Humanities Institute Seminar on "Interface" at Duke University, a yearlong examination of relationships between humans and machines, from the pre-Socratics to Virtual Reality. Co-convened by Professors Timothy Lenoir and Priscilla Wald, our Seminar has met weekly all year, and brings together sixteen faculty members, graduate students, postdoctoral fellows, librarians, and technology innovators for weekly conversations. The Fellows in the Interface Seminar served as referees for the scholarly papers at the conference and will serve as session chairs and commentators throughout the conference.

Finally, **WELCOME** to the latest in the John D. and Catherine T. MacArthur Foundation's series of regional public events on digital media and learning. HASTAC is very proud to participate in the MacArthur Foundation's exciting new initiative on Digital Media, Learning, and Education.

* * *

If six different welcomes are required, that's indicative of the kind of interlocking, networked projects sponsored by HASTAC. Nothing HASTAC does is linear. We are a network, not a traditional organization. We believe in what Tim O'Reilly calls "Web 2.0," this generation of internet interactivity, social networking, customization, and collaboration. HASTAC may well be the first "virtual university," a university without walls, departments, or even traditional disciplines, and we see our charge as making a contribution to lifelong learning, from cradle to grave. We believe that specialized academic knowledge should be put at the service of society at large. David Theo Goldberg and I came up with the idea in 2002, and since 2003 HASTAC has been meeting twice a year, supporting seminars and workshops, new courses, new programs, developing new technologies, forging new collaborations, working in communities, and, in other ways, combining our efforts and our ideas in order to promote innovative and collaborative models of thinking. We are entirely voluntary--an international knowledge network of educators and digital visionaries committed to the creative use and critical understanding of new technologies in life, learning, and society. Anyone can join, simply by registering on the HASTAC website, posting news releases or job openings or collaborative projects in our forum, or contributing ideas by using our open blog tools. You can exchange information there, find partners, and become as involved as you want to be. Or you can just go onto the site whenever you wish and find out what's happening. Our rules are minimal. Creative Commons licensing, and internet collaborative civility rule. Our goals are huge:

We are convinced that, if we work together, if we pool our resources, our ideas, our imagination, our skills, and our technologies, we can transform the Information Age into an Age of Understanding.

I know that sounds good, but what does it mean? Let me give you a brief example. I've been fortunate to be a fellow this year in our Interface Seminar at the John Hope Franklin Humanities Institute. As a fellow, I am released from half of my teaching responsibilities for the year in order to learn a new topic that I can then bring back to my future teaching and research. Even better, I'm in the Seminar with seventeen other faculty, librarians, technology specialists, postdoctoral and doctoral students. We meet every week to discuss our evolving ideas together. For my project this year, I've had the extraordinary privilege to spend my time reading about infant knowledge acquisition—how do humans learn how to understand the world. Since I'm not a neuroscientist, I have the flexibility to read everything and anything. It may sound like a very narrow area but, in fact, there are thousands of experiments, studies, articles, and books about how infants develop what philosopher Elizabeth Grosz calls "concepts." What I've found is that highly-trained specialists in one field often do not read in other fields. And very few in this area of infant cognition are aware of the whole human history (in all cultures) of thinkers who have analyzed the very nature of what a "concept" is. So a brilliant and creative lab experiment will often end with a generalization about the human mind that, from the long view of the humanities, is nothing short of sophomoric. By the age of fourteen months, one experiment shows, an infant can recognize members of its own "race" but the experimenter's definition of "race" is so simplistic it is contradicted by the evidence he presents and thus undermines the significance of the experiment itself. But how could it be otherwise? Why should this superb cognitive psychologist be expected to know critical race theory? It would be like asking the person who engineered a high-performance car engine to also be responsible for designing the car's beautiful and aerodynamic exterior. If BMW has figured out the fine art of collaboration, why can't *educators*?

That is why we are here for the first international HASTAC conference. Since the late nineteenth-century, education has emphasized specialization, to the point that we now live in an era of knowledge *segregation*. But that has to change. There has never been a great age of science, in the history of any culture, without a coterminous flourishing of the arts and the humanities. The reason is obvious. New technologies change us. The brain is a learning machine and, with each technological development, we have different tools, different information, different transportation or communication that changes relationships between our minds and bodies and that of others. The substance that we learn changes how and what we learn. But it is the lessons of history, theory, philosophy, literature, and the arts that allow us to frame these changes, to categorize and conceptualize processes too minute and too extensive without such containment. Thomas Kuhn famously called this a "paradigm shift." If science and technology create conditions by which a paradigm shift might occur, it is historians, philosophers, and artists who gather the discrete threads of influence and difference and make them into a narrative, a coherent and connected set of practices that mark and demarcate change. *That's* the paradigm shift.

The Information Age is too important, too revolutionary, to be left solely to scientists. We need to capture this moment, understand it, exploit it, push it, make the most of it. The Information Age is too important for knowledge fragmentation. Without a concomitant shift in intellectual paradigms, we cannot make the vast scientific changes of our era *meaning-full*.

That is why we have gathered here a group of highly trained specialists who understand the need for rigorous collaboration. In this era of cutbacks to education, funding increasingly comes from national and private philanthropic agencies. Often these funding agencies also have segregated functions that reify the compartmentalizations of knowledge and, indeed, encourage one discipline or division to compete against the others for funding. So we are also fortunate enough to have at this conference top thinkers from many of the most important of these agencies who will help us theorize how we might move beyond the moment's intellectual fragmentations. More practically, we hope, these funders will also have the opportunity to think of ways that they might collaborate between and across agencies to set new models for the universities who so depend upon their largess for support. We all need to work together to rethink the interfaces of fields and institutions.

We are contemplating the future of learning together--and we thank you for being part of this very exciting journey. If you look to your right and to your left, in front of you and behind you, you will see seminal thinkers from virtually every field of knowledge, from the academy and foundations, from K-12 education to policy makers. This is a very rare moment, an interface moment. Nothing quite like this has ever happened before. But, I promise you, on behalf of the HASTAC leadership group, that it *will happen* again. And again. And thus our motto for this gathering: "The future is somewhere *here*."

Funding the Digital Future*

Leaders from national agencies, private foundations, and industry discuss digital funding opportunities, initiatives, and visions.

Brett Bobley, CIO and Director of Digital Humanities Initiative, National Endowment for the Humanities

Karl Brown, Associate Director, Applied Technology, The Rockefeller Foundation

Jerry Heneghan, CEO, Virtual Heroes, and Chairman, North Carolina Association for Advanced Learning Technologies (NCALTA)

Gary Kebbel, Journalism Program Officer, John S. and James L. Knight Foundation

Kevin M. Guthrie, President, Ithaka; Diana Rhoten, Program Director, Office of Cyberinfrastructure, National Science Foundation

Steven C. Wheatley, Vice President, American Council of Learned Societies

Constance M. Yowell, Director for Digital Media, Learning, and Education, John D. and Catherine T. MacArthur Foundation

* The audio for this panel is available at www.hastac.org.

Interface Genealogies

Previous moments in media history illuminate what is and isn't new about "New Media."

Caitlin Fisher
"Interface Epistemology: Hypermedia Work in the Academy"

Lisa Gitelman*
"Xerographers of the Mind: The Lost Idea of the Photocopy"

Matthew Tinkcom
"Eduction: A Theory of Value in the Digital Cinematic Epoch"

* Available in digital format in the video archives of the conference at www.hastac.org.

Interface Epistemology: Hypermedia Work in the Academy

Caitlin Fisher

It's a curious thing to find that a project close to your heart is now historical, but a decade ago I embarked upon a doctoral dissertation in and about hypertext and even saying that word sounds about a million years old, so there you have it.

I want to tell you a bit about the story of writing and reading that project – a native hypermedia work completed at York University in 2000 just as that institution was circulating a discussion paper proposing that all electronic dissertations be submitted with 12 point Times Roman font and one-and-a-half inch margins regardless: the future of writing as pdf. Then, as now, I saw the future of writing somewhat differently: I was particularly interested in the epistemological status of interface, especially the capacity of interfaces to make connections and arguments intelligible to readers.

My dissertation, *Building Feminist Theory: Hypertextual Heuristics*, was an exploration, in hypertext, of the resonances and productive couplings between digital writing technologies and feminist theories. Implicit in the title was the claim that the process of shaping hypertext was itself a form of feminist theory production. Rather than simply identifying feminist hypertexts and explaining them in terms of a feminist hermeneutic, the dissertation used theory to build a new kind of text, a text which sought a form resonant with the bordercrossing narratives and subaltern knowledges it sought to explore. Understanding the interface and the text to be co-constitutive of meaning, then, I struggled at all stages with the choice of interface and with the limitations of code available to me at the time of writing. I'm going to discuss and show you some examples of how the interface to the work evolved, possibly regressed, and, is now, I hope, evolving again.

At the time, I was also starting to read and write hypertext fiction and I loved Deena Larsen's account of how she began writing hypertext:

I wrote a series of stories about women in a Colorado mining town. But the stories weren't enough to show the relationships some too secret for words, some the characters didn't understand. So I put the stories in little houses on a model train set and strung different colored embroidery thread to show different connections. But you could not follow the two-inch thick maze of thread. My friends were very supportive. Comments included: 'You idiot. Do this on a computer. Here is how to work HyperCard. Now get this thing out of my basement.'[1]

I stuck that quotation on my computer and thought it was a great model for academic work in social and political theory. (Yes, it took me a long time to finds a supervisor ;-) But long before I was introduced to hypertext software I, too, wrote hypertextually. I have always written this way: with straws and string and handwritten letters; cross-legged on the floor with my scissors and glue-stick; in an empty room arranging and rearranging four hundred eight 1/2 by 11 handwritten sheets, drawing arrows with thick magic markers (that was my Master's thesis covering an entire room and, yes, it did make me crazy). Like my feeling of 'coming home' to feminism, a concept problematized by now, I experienced digital writing in hypertext as yet another homecoming. To me, putting these two things together – hypertext writing strategies and feminist theories – seemed obvious and irresistible. I was curious, then, to find when I embarked upon the project in the late 1990s that, for the most part, the emerging hypertext theory paid little attention to either women hypertext practitioners, or feminist theories that had come before. The literature, as Burnett also observed at the time, was "conspicuous in its omission of female writers and feminist critics, not to mention writers of color."[2]

And since this panel is in part about interrogating what might not be so new in new media I'll add that this absence of critical attention in the literature was particularly striking given that feminists have always had a particular investment in the creation of new genres and structures and breaking the mold of narrative forms. Feminist theory has a long history of putting forward claims regarding what is at stake in adopting new ways of thinking, storytelling, writing the unspeakable, theorizing empty spaces and absences. As Sara Diamond notes, "the circular pathway, for example, has long been considered a feminist quest myth" remarking, further, that many of the features struggled against in feminist experimental practice - "inciting incidents, narrative peaks, troughs and closure"[3] were, in hypertext, already absent.

When I began to conceptualize my doctoral project, I assumed initially that I would look at a handful of hypertexts that I could argue were radical or otherwise interesting in terms of narrative strategies and use of code, make claims for their 'feminism' and then allow these works to guide my exploration of the 'newness' of the medium. The process of actually working with these texts, however – and my own experience of encountering hypertext as familiar, as a continuation of my own writing practices – forced a reconsideration of my investment in theorizing hypertext as presenting anything like a radical rupture with experimental practices over the last century. I opted instead to situate feminist hypertext in terms of its continuity with other experimental writing and visual practices and as my definition became more fluid, the dissertation grew to accommodate many different texts, practices and images, many of them not contemporary or digital at all. And this wide-ranging interdisciplinary approach to how literary and argumentative hypertext might be brought together in conversation, how images and text work in a hypertext environment, how feminist theories of print and film, autobiography and critifiction, desire and social difference might engage hypermedia practice resonated strongly with an interest in boundary-crossing evident in both hypertext and feminist scholarship.

I brought other preoccupations to the project, too, of course: my interest in ephemera, the marginal and the relationship of collecting bits and pieces to identity; Walter Benjamin's Arcades project; McLuhan; a fascination with quilting and piecing fabrics and stories, digital stitching together of all kinds; building blocks and the Eames houses of cards; collage work and mash-up.

There are connections here, too, with Matthew's work on amateur prosumer culture and the autobiographical impulse. Almost all of the images I made or adapted referred directly back to the lessons suggested to me by the hypertexts themselves or the feminist theory to which I turned initially to help me to understand these practices. The autobiographical, first-person impulse of these early works was striking and, in the end, images included in the dissertation were, with few exceptions, my own – the exceptions mostly being images taken/stolen directly from the hypertexts that inspired, informed and shaped the work. The pages were in many ways self-portraits, or images of friends, hand to the scanner. I also scanned pieces of fabric, bits of quilting, braids cut off and saved in tissue, petals saved for years between pages of favourite books. And in keeping with the collaborative spirit and collective memory-making enterprise of some feminist hypertexts, the dissertation is also a mnemonic system, filled with small gifts from people who encountered the text along the way to its completion: laughing into my microphone so I could make .wav files, film from an MRI so I could have images of spines and organs, baby pictures, pockets turned inside out to reveal mysterious things to put under my digital microscope . . .

In the end, *Building Feminist Theory: Hypertextual Heuristics* consisted of over 1400 lexias. But the intellectual core of the project, and the most interesting aspect of hypertextual writing to my mind, was the constellation of ideas held aloft by the technology. In the case of this dissertation, the web of original lexias, quotations, imagery and sound put into conversation was held together by more than 17,000 links. In this way the linking structure was the intellectual core of the project. Indeed, the linking structure – the ability of this writing technology to hold the all-at-onceness of theory as we build it, to communicate this constellation of ideas, and crucially, to have readers encounter and explore them (though never unmediated, of course) – is, potentially, the most theoretically interesting aspect of hypermedia writing.

As theoreticians, we are, of course, used to reading across texts and complex arguments. And these associative webs we develop as readers are intricate. What we are less accustomed to encountering is the way others read across texts and complex arguments before these are communicated to us as single, if complex, narratives for academic publication. As I look back on the production of the dissertation and the way I hoped it would communicate, I still see the promise of hypertextual writing as allowing me to deliver on disk, an encounter with, at once, my library, my theoretical orientation, the way I made (sometimes contradictory) sense of (these) texts, and understand myself to be positioned by them, my reading and visual art practices, the themes that recur as I come to new understandings, through new encounters and re-readings and how I come to encounter and generate new knowledges. It felt and still feels very de Certeau in the sense that "exhibition, showing, making visible" is understood as a form of analysis and of theory-making. And the process of constructing the work certainly made me feel like I had an insider's intimacy with Adorno's Negative Dialectic. Ultimately, I associate the hypertext with the scaffolding of the academic enterprise, the unconscious of the philosophical line, whose communication, I offer, has real academic, theoretical and aesthetic value, namely: the concretization of a web of signification.

And so you can imagine my surprise – and disappointment (ok, horror) – when the piece began to circulate beyond my committee, the key interlocutors to this work, in its final form – html – and that's not how the piece seemed to function.

One of the first things I came to know when I began to share my dissertation widely with readers was that, more often than not, my readers read nodes and not links. These reluctant bricoleurs read words and quotations only, and so the lexias were understood as the content of the dissertation and the structure itself – basically what constituted the epistemological intervention – its contours, its conventions, any new ground I'd hoped to break – was largely unintelligible to many of them. For some months I understood the work as a catalogue of losses – the loss of polemic, of certain kinds of rhetorical gestures, and of mastery for which I was prepared (this being hypertext, after all) but also the loss of a community of like-minded thinkers with whom to share the project. While I believe new literacies are emerging and even ten years ago we were talking about a new grammar, aesthetics and poetics of digital texts, I had undertheorized the ways in which readers – expert readers of linear texts – would experience hypertextual work at that time.

Soon after I defended, I attended a conference at which Diane Greco, then acquisition editor at Eastgate, proclaimed that the urgent call to readers was "to learn to read archives."[4] And this sounded just right to me, given that something that's 1400 anything long is an archive and a linking structure in itself and doesn't teach anyone to feel comfortable crossing it. I love that this panel is also talking about Xerox and archives, then, both things being close to my heart while working on this project because essentially what I had produced was a large archive with instructions for reading.

Just about that time, too, Lev Manovich declared the database to be the new symbolic form of the computer age and I was intrigued by that, of course, and also with his idea that the specificity of new media might well reside precisely in the way that the database can be layered with multiple interfaces – and that being able to read datasets and collections has, of course, been commonplace in different historical periods. "For centuries," Manovich writes, "a spatialized narrative where all images appear simultaneously dominated European visual culture then it was relegated to 'minor' cultural forms as comics or technical illustration. 'Real' culture of the 20th century came to speak in linear chains, aligning itself with the assembly line of an industrial society and the Turing machine of a post-industrial era. Two competing imaginations, two basic creative impulses, two essential responses to the world."[5]

In response to these ideas, I began to revisit the way my project had evolved. *Hypertextual Heuristics* was first conceptualized as spatial narrative and built using Storyspace – software that enables a spatial layout of information. It was exported to html at the last minute because, at the time, Storyspace was pretty clunky in its handling of rich media. The conversion seemed like a good idea, especially given the potential ease of dissemination of the work over the internet. In retrospect, however, moving from Storyspace to html was a mistake. In html, the work performed like a linear catalogue. What I hadn't bargained for was the way the export flattened out the visual cues given to readers about the relationships of texts to each other and forced a primacy of hierarchical relationships on a text that had not been coded for in the beginning. Listening to Greco and reading Manovich I was struck by the realization that the dissertation functioned problematically because the spatial aspects of my work were no longer apparent and the cues for reading the archive were gone.

In performing the translation from Storyspace to html I had evacuated the spatial dimensions of the project and, in Manovich's terms, aligned myself with the assembly line. This was especially vexing because I had, even before embarking on the project, seen my texts as three dimensional and sculptural – as a thought sculpture. But it was clear that the interface I had chosen meant that the sculptural form of the argument was not intelligible to readers, the contours of both the archive and the argument I was making about it were lost, and the most interesting feature of the writing – the constellation of ideas held aloft by the technology through its linking structure – failed to communicate. Worse, for a piece that argued for the navigational apparatus itself as a signifying component of the text, the html version worked against the kind of knowledge I was trying to produce. So let me just show you a bit of this.

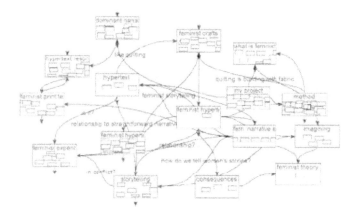

Figure 1: An early screenshot of the dissertation. Spatial, but two-dimensional, a point to which I will return. (An aside: the Storyspace datasets have all begun to disappear… I was going to show the early interface in action for the conference, but the disks are actually erasing themselves from my disks like missives from special agents!)

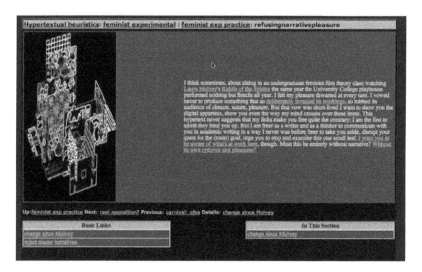

Figure 2: Here are some screen shots from the html version. Note that the menu bars at the top replace the arrows drawn in the Storyspace programme. This isn't simply a design issue: it's an epistemological issue. A struggle with

interface is a struggle with meaning and knowledge production. I'm reminded here of Walter Benjamin: "for the important thing to the remembering author is not what he experienced, but the weaving of his memory, the Penelope work of recollection."[6] In considering what is at stake in theorizing interface epistemology, then, we need to ask how we can make intelligible the linked and coded concretization of the weaver's constellation.

My understanding of the constellation and its philosophical and political importance emerges from my reading of the Frankfurt school. When we want to understand an object of interest – in the case of my doctoral work, for example, feminist hypertext theory – we must not look directly at the object, fetishizing the concept. For Benjamin, the constellation is a multidimensional form: the arrangement or configuration in which a variety of concepts, models, ideas or other materials takes shape. In Adorno's extension of the idea, the constellation is addressed this way: "as a constellation, theoretical thought circles the concept it would like to unseal hoping that it may fly open like the lock of a well-guarded safe deposit box: in response not to a single key or a single number, but to a combination of numbers."[7] McLuhan's prologue to Gutenberg galaxy echoes the idea and approach: "the galaxy or constellation of events upon which the present study concentrates is itself a mosaic of perpetually interacting forms."

I use the word constellation with a nod to Benjamin, Adorno and McLuhan among others, then, but I suggest it's different here in new media – how? Because, crucially, this particular constellation has been coded, because the linking structure is saved in computer memory, memory beyond my own, I can return to it, and share it with you, I believe, if only I can find better ways of transferring this cat's cradle to your hands, without this act of communication collapsing into a puddle of string, or html, or even 2-dimensional Storyspace.

Ten years on, we're already better at reading new shapes and we are learning to read archives in new ways. It's hard to imagine anyone now, for example, who would need assistance in following a web-based link. But finding the right kind of interface for the kind of thought sculptures we make as theorists continues to present challenges. My current work in augmented reality interfaces and storyworlds, while seemingly a long way away from *Building Feminist Theory*, has actually drawn me back to this early work, to re-examine its possibilities and poetics. I'm working to port it back from the html into a form more suggestive of the constellation I'd like to share… maybe add a dash more pleasure and give up a bit of jouissance. I'm working with Thinkmap software – a commercial solution that allows for the construction of a dynamic 3-dimensional interface to the archive I've built, to arrive at an interface that will at least suggest to readers a sense of breadth, the argument I make for the necessity of a shifting conceptual center to the work, and a way of navigating the piece that invites an understanding of the arguments about interrelationships I try to make. I'm also thinking of porting the whole experience to augmented reality, in the end… a full body thinkmap interface where 1400 nodes can hang like stars, sewn together with virtual string, ready for walk-through.

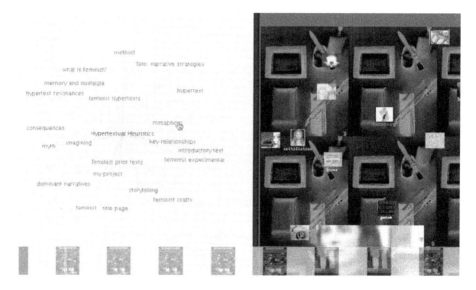

Figure 3: Thinkmap screenshot

What is at stake more generally with respect to knowledge and theory-production here? Which of our interfaces are successes? Failures? For early hypertext practitioner Michael Joyce the litmus test of any hypertext was whether it allowed its users to look at knowledge in new ways.[8] To that I would simply add that working mindfully at the interface gives us new tools to build knowledge, too, to craft knowledge in new ways. Even as my own early work risked intelligibility, this labour was well worth the risk.

Endnotes

[1] Ley, Jennifer. "So You Thought You Understood Hypertext?" *Riding the Meridian* 1.2 (1999).

[2] Burnett, Kathleen. "Toward a Theory of Hypertextual Design." *Postmodern Culture* 3.2 (1993).

[3] Diamond, Sara. "Gender and Technology," *Variant* 16, Winter/Spring (1994): 65.

[4] Greco, Diane. Comments made during Hypertext 2001, Århus, Denmark, August 14-18th, 2001.

[5] Manovich, Lev. "Database as a Symbolic Form," *Millennium Film Journal* No. 34 (Fall 1999).

[6] Benjamin, Walter. "On the Image of Proust" in *Walter Benjamin: Selected Writings, Volume 2, part 1, 1927-1930* Edited by Michael W. Jennings, Howard Eiland, and Gary Smith.

[7] Adorno, Theodore. *Negative Dialectics*. Continuum, 1973: 162-163.

[8] Joyce, Michael. "Siren Shapes: Exploratory and Constructive Hypertexts," Michael Joyce, "Siren Shapes: Exploratory and Constructive Hypertext," *Academic Computing* 3: 12.

Eduction: A Theory of Value in the Digital Cinematic Epoch

Matthew Tinkcom

I. Introduction

Prior to the emergence of digital modes for the storage, circulation and exhibition of moving images, critical commentaries on the uses of non-professional film and video production emphasized the seeming "dead end" of these media for the production of historical knowledges outside those officially sanctioned by most scholars. Indeed, in a terrible epistemological paradox, the seeming amnesia around these "amateur" modes fostered the disappearance of these very same media; writing in 1995, Patricia Zimmerman commented that the potential loss of these materials—both in their physical destruction by neglect and in their absence from film canons—is "not simply an inert designation of inferior film practice and ideology but rather is a historical process of social control over representation."[1] Taking up this challenge—the challenge of recuperating amateur film for new understandings of such phenomena as nationalism, domesticity, and queer identities—Jane Simon argues that digital media offer significant remediations that complicate such forms of social control, but she cautions us not to be utopian in embracing these technologies simply because they are thought to be "*new* media." In her discussion of the transfer and bundling of 8mm amateur cinema in Australia to the dvd format, heralded by the *Homemade History* series about amateur film in Australia from 1950s and 60s, Simon argues that the "positioning of digital media as a discontinuity with obsolete forms of media reveals a desire to avoid stabilizing the new and the future as a mere continuation of the past".[2] Put another way, the adaptive re-use of older media in digital forms may not be as disruptive of what we want from or for the "past" if we continue to write the same historical narratives and see them as confirmed by what we apparently have discovered, through digitalization, of older non-canonical media.

At the same time, Simon offers the sense that digital remediations of prior cultural forms *do* make possible new practices and social relations that need to be seriously engaged—were it not possible, for example, to screen in the *Homemade History* dvd Ken Garrahy's super-8 footage of gay and lesbian social clubs in the 1960s, we would be ignorant of the presence of queer social lives in Australia in the pre-liberation politics historical epoch, and in this regard what is made possible is what Simon describes as "histories of *particularity*" [emphasis in the original] that can be accessed "without the difficulties of accessing either home movies or projectors" and "that can be easily reproduced, distributed and accessed through libraries." (Worth mentioning, as a way of bringing *particularity* to this digital mode of historical knowledge production, is the sense that not all libraries, at least in the U.S., would be amenable to making available queer alternative historical media—at least where I live.)

Frank Gray and Elaine Sheppard, creators of an on-line resource to regional digital moving-image archives in the United Kingdom called *Moving History*, note that the practices of official digital archivists shape our larger notion of the film canon in unexpected ways—and, indeed seem to honor an evaluative scheme that is inadvertently quite the opposite of what Zimmerman discussed ten years ago. According to Gray and Sheppard, because public digital archives honor intellectual property rights and cannot post copyrighted materials—and frequently do not have the resources to pay for royalties—the bulk of their content is derived from "orphaned" materials that are transferred to digitalized forms. They write that "the project's duration and funding did not provide for either copyright license payments or film-to-video transfer costs. These restrictions unfortunately meant that some potential material was excluded automatically. However, these conditions also meant that in some cases less prominent and less-used moving images were selected and allowed to come to light through the website."[3] The activities of the archivist—cataloging, digitizing, publishing to the world-wide-web—not only make previously unseen cinema available to new users, but allow us to screen films from widely dispersed archives in relation to each other—for example, scholars working on gender, consumerism and fashion would be well-served to screen both "Fashions of '38," a 1938 8 mm silent film that depicts home-produced fashion shows[4] and "Risqué Dresses," a 1970 news piece on the appearance of décolletage in the southwest of England.[5]

If the changes described by these scholars and archivists are acutely felt and theorized at the institutional level, let us consider the effects they have on those who use the archives for their own purposes: to manufacture moving images in the digital mode.

II. Eduction: The Archive, History, Value

In an essay published in 2006 in *The New York Times* called "The Secret History," Herbert Muschamp offered his own assessment of the decade long controversy over the refurbishment of Manhattan's Number 2 Columbus Circle. Designed by Edward Durrell Stone in 1964 and brandishing a highly ornamented neo-Venetian Gothic marble façade, the building was, in the moment it was erected, highly at odds with the particular glass and steel modernism favored in the post-World War II United States. Until its refurbishment last year, in the preceding ten years the building has been the object of a pugnacious debate brought about by developers' attempts to strip it of what Muschamp calls its "first lady architecture." Muschamp argued that the Stone building should not be touched. Strikingly, though, he does not defend its particular style. Rather, he argues that the building should be preserved in all its anomalous glory because of its status as a part of New York City's history as a gendered space. The Stone building was commissioned by A&P grocery chain heir Huntington Hartford for his collection of pre- Raphaelite, impressionist and surreal art. Until its closing in 1969, it was a space where many queers met, drank, ate, and cruised in its penthouse restaurant, the flamboyant tiki-themed "Gaugin Room."

Muschamp's campaign to preserve the Stone Building in all its ornate architectural glory was, at base, a question of archives and a question of value. Muschamp, in essence, argued that Number 2 Columbus Circle should have been preserved because it changed the canon of architectural value championed by the city's Landmark Preservation Commission. The Commission had remained silent over the possibility of a queer history residing, literally, in the walls of some of Manhattan's built space—as if aesthetics and preservation (that is, the archive) can be separated from issues of value and social history. Muschamp "outted" 2 Columbus Circle and, by implication, suggested that preservation is never innocent or disinterested in its shaping of the uses of history. He indicted the Commission for its failure to protect this building, and ended his commentary with a polemical flourish: "A vibrant city is perpetually recreated from the emotional depths, and from our socialized capacity to empathize with the memories of others. A landmarks commission embodies this capacity in administrative form. It should be the agency's business to know when somebody's memory is being stepped on."[6]

I open the question of history, queerness and value in this essay in terms of the idea that the metropolitan landscape might be filled with instances of queer memory and thus functions as a potential archive because I think that it poignantly has implications for the even more fragile media of the moving image. I will argue that the preservation of a queer archive, and who seeks to preserve it, raises questions about the nature of how those parts of our lives— the private arena to which sexuality is too often consigned in our habits of thought and the public sphere where debates about what is to be preserved occur—are more bound up in each other than our labors may realize. I will further argue that recent media technologies have implications for the expansion of this archive into the habits of *production* and *viewing*—both—of cinema. So, I am arguing that the archive is not defined solely as those texts, objects, images, sounds or buildings which are understood as worthy of preservation but includes also the *impulse* to archive, and that this impulse conjoins the aesthetics of use to the aesthetics of materiality. Muschamp charged the New York City Landmarks Commission to understand the city's counter-memories as part of New York's public, municipal identity, and this re-shapes the archive as a very queer thing.

As Jacques Derrida notes about the archive, "it shelters itself from this memory which it shelters: which comes down to saying also that it forgets it."[7]

For Derrida, the archive's status as an official form of memory which is preserved for the sake of power undercuts and is undercut by the sense that something which antagonizes sanctified memory is contained paradoxically within the very archive itself. As Derrida notes, the archive was conceived in Greek culture as a domestic activity, one where the larger political world was prohibited from tampering with the contents of the household, and thus the private home protected memory from revision. By contrast, the queer counterhistorical archive of the motion picture—evanescent, increasingly mutable, especially in its digital form—presents us with archival issues every bit as political and infinitely more malleable—the home is now the space where so many non-professional users revise the archive. As I have argued previously (in my book on camp), consumption *and* production must be theorized as entwined as we consider the queer archive of the cinema. To make visible what is hidden in the archive is, in a sense, to queer it, or at the very least to recognize that those memories seized upon by the archival impulse move across a threshold similar to that of queer value: they lose the very power of their anonymity as they become more publicly recognizable.

I find the analogy between urban space and the moving image useful for understanding the transfer of cinema to its specific digital iterations because I am interested in the astonishing mutability of filmic archives, and the relative instabilities of all of those categories within film history, criticism and theory. The transfer of cinema to its specific digital iterations that is occurring presently destabilizes the idea of film and its archive—indeed, the idea of film *as* an archive—to a such a remarkable degree that we should take pains to understand the effects of this process upon some of our most central assumptions about the nature of the cinematic text. Most historians and theorists of the cinema have been able to take advantage of the sense that, whatever other effects a film might have upon its audience, it was "finalized" as an object of exchange within the economic chain at the moment of its exhibition; to use Marx's locution, its "material substratum" was complete. The rich body of scholarly work of the past three decades on the topic of spectatorship has sought to demonstrate how reception manufactures its own versions of the film text—through fantasy, pleasure and displeasure, through film as social text—but these arguments, too, were predicated upon the idea that the commodity of the film was highly stable as an economic object.

In comparison, the present historical epoch marks an astonishing displacement of film's manufacture into the sphere of consumption, and with it arrive new opportunities (or demands, depending on how you see it) for cinema's "users." This shift has occurred through what economists would note is the migration of several factors of production across the production/consumption divide: the machinery for the manufacture of the moving image has taken new shapes in small, comparably affordable "prosumer" cameras, film-stock is increasingly supplanted by digital recording modes, and editing, visual and acoustic effects become markedly standardized and professionalized through pc desk-top software applications. We need to attend carefully to our moment's displacement of production into consumption, borrowing from what the marketing nomenclature of "high-end" consumer technology calls "prosumer" uses. Whether in, for example, Andy Warhol's films of the 1960s or the Stone building, "prosumption" has been with us a long time, even if it has not been as visible in past eras as it is in our digital moment. Now, however, prosumption is everywhere. Prosumption *is* key to the digital moment, but precisely how, and why prosumption operates is the question of value—and of queerness—that I am addressing. And, I am here making the link between Andy Warhol's cinema and the preservation of the Stone building because, in my mind, it is not simply a coincidence that Andy Warhol's cinema did much the same thing that we are being asked by Muschamp to do: to "read" a record of both the queer lives that were lived there, and as a *queer* record of how to read those lives. Warhol himself was perhaps the original "prosumer" because his cinema and art continually seized upon financially accessible forms of mass media—16mm film stock or silkscreening—to produce from such consumer technologies highly "professional" works of art.

In our own moment, "prosumption" derives from the digital modes' ability to reproduce both the cultural commodity and the modes of its production with high fidelity. Its economics are tricky. When someone says, as is the case of the film I discuss below, *Tarnation*, that a movie was made for $300, what do they mean? And what does that bargain basement pricetag mean for the film industry writ large? Put another way, in more realistic and more historical economic terms, the last two decade's intense and virtually incalculable capital investments in technology, and especially software development for industrial applications, is the foundation of every three hundred dollar movie. Yet the production values of current digital cinema—and the sophisticated labor of those who have mastered or been weaned as youthful consumers on digital technology—now presses the boundaries of the film industry itself and competes with that formation. The capital investments in technology of the last decades are derealized and that capital is diluted across myriad computer hard-drives where the capacities for handling image and sound through new interfaces facilitate significantly more individuated modes of production.

How the archive is understood, indeed what constitutes an archive, and the kinds of meanings that might emerge from a variety of archives—official, personal, secret, self-made—matters immensely to the question of digital cinema because the transfer of the archive into the larger matrix of digitalized image culture has immense economic, social, formal and ideological repercussions for the ways that we think about the moving image in its past and present moments and, of especial significance for this project, about who seeks to work upon that archive. This argument is reminiscent of that in *Working Like a Homosexual:* in camp, subjects who for such a long time have found themselves distanced from the sphere of production—situated within strict prohibitions upon those representations related to dissident sexualities—found ways to make use of the techniques of cinema—in the use of color film stock, in editing, in the varieties of performance—in order to create a new moment in cinematic culture. Thus I am arguing that one way of seeing the new labors of digital cultural forms—*prosumption*—is by analogy to this history of camp part of a larger historical tendency in which more longstanding media are liquidated by *particularized* forms of labor into digital forms and in this process they are renewed as possible commodities. Note that I have said "possible." I insert that conditionality because part of the problem at hand is that digital cultural forms are not required to behave as objects of economic exchange, even as they experience a re-valuation in various social and aesthetic registers.

Indeed, I would argue that cinema's transformation by the new digital media of software design, dvd-formatting, digital storage modes like Tivo, and the world-wide web occurs most centrally in terms of its relation to prior instances of recorded visual, acoustic and textual culture—the archive—and that the digital formats of the archive shift many of the *activities* of production away from the *sphere* of production. And, they do this at the moment in which the digital money form and digital cultural production find themselves as similar information forms in the nexus of exchange more intensely than in any prior historical epoch.

This new nexus of exchange is such a remarkable change in the nature of cultural commodities that it calls for a different theoretical category by which we might make sense of the labors involved. I am calling this form of labor "eduction," a process defined by *the attempt to extract new forms of value from the material substratum of the digitally preserved cultural product*. I am arguing that, while there are prior historical analogies—such as Warhol's invocations of studio star glamour or Kenneth Anger's readings of Hollywood gossip—digital media are more dependent on eduction than cinematic forms of the past—in scale, in scope, in potential, *and* in value.

I'm devising the term of eduction from the latinate verb "to educe," which in its multiple meanings can suggest, variously, the activities of drawing out something hidden, latent, or reserved; branching out such as a river or blood vessel might do; evoking or giving rise to a new version of something. The common features of those things that are educed are (1) that they call for the refashioning of multimedia aggregates into new *narrative* forms (2) that they contribute to the *attribution* of values (be they historical, ethical, aesthetic, affective) but not necessarily the *extraction* of subsequent economic value (3) that eductions participate in preservation but also in the revision of older cultural productions in their transfer to digital forms and, finally, (4) that the economic implications of digitalization means that those texts that are produced through the extraction of new value frequently "short circuit" the economy by moving through networks other than those of the marketplace.[8]

III. Jonathan Caouette's *Tarnation*

Eductions take many forms. Among them: the redeployment of the back-catalog of Classical Hollywood film in the media channels of Turner Classic Movies and Netflix, the use of 1960s martial-arts cinema in the hyper-edited and hip-hop styled cable program *Kung Faux*, the deployment of film-as-wallpaper in the mise-en-scène of Steven Spielberg's 2002 feature, *Minority Report*, or the redistribution of web-based prank films on Bravo TV's program, *Viral Video*. Non-industrial, more personalized eductive endeavors might include Joan Braderman's feminist videos such as "Joan Does Dynasty," Tom Joslin and Peter Friedman's *Silverlake Life*, Joe Gibbons' *Barbies Auditions*, and Todd Haynes's *Superstar*. Here, I want to examine a more recent production that exemplifies what I have been arguing about queerness, the archive, prosumption, and eduction: Jonathan Caouette's 2004 release *Tarnation*.

In an opening sequence from *Tarnation*, Jonathan Caouette fashions a home-movie gone terribly wrong. Sending up the conventions of the kinds of sentimentalized home video compilations that are frequently made to commemorate anniversaries and birthdays, Caouette offers us an account of how his mother's life, and his own, lost direction and developed into a nightmare at the hands of doctors, clinics, foster homes and other family members. Variously a documentary, a multimedia self-portrait, and a cinematic poem, *Tarnation* is, not least, a tribute by Caouette to his mother, Renee, and the harrowing effects of a lifetime of medical treatment on her physical and mental wellbeing. The film additionally offers an account of Caouette's development as a queer man and an artist by editing together his personal archive of family photographs, super-8 home movies, video footage, audiotapes, answering machine messages and popular music, all processed in Apple iMovie software. More specifically, *Tarnation* offers an almost singular instance in film culture in which a "perverse" subject is able to construct these narratives of self and family through the use of such diverse media, and the film is particularly significant in light of its refusal to adhere to the more commonly available narratives in which queer subjects are understood to mature within the violent antagonisms towards same-sex desiring people. Indeed, even within the bulk of seemingly well-intentioned accounts that emerge from the sense that lesbian, gay, bisexual and transgender people, like everyone else, have families and that their families participate in the maturation of queer lives, there is frequently the sense that no queer person can speak on his or her own behalf, much less offer, as Caouette does, a story that seeks to explain not solely how he or she "got that way" but how it is that being "that way" is perhaps one of the few possible ameliorating effects of the traumas visited on his family by years of Renee's electric shock treatments and the use of prescription psychotropic drugs. *Tarnation* appears as one of the few accounts to consider positively "the huge blank spaces in what purport to be developmental accounts of proto-gay children," as Eve Sedgwick describes them, that circulate more generally within the literatures of psychology and psychoanalysis.

If understood as the effort by a queer son to invent a genealogy for himself, *Tarnation*'s status as an archival project stems both from its ability to recuperate the detritus of everyday recorded life and to situate those materials in relation to a more expansive history of queer self-representation. That is, it educes queerness from its media sources. The personal archive, which surely expands with every voice-mail, electronic communication, digital image and blog entry we ourselves might make, is within Caouette's account framed as part of the queer avant-garde life that he shaped for himself while growing up in the context of Renee's repeated hospitalizations and the life he led with his grandparents and in foster homes. This genealogical project *educes* both the recorded details of his own life and the political and aesthetic dimensions of the avant-garde that, in a sense, helped to raise Caouette if not indeed preserve him from the familial disarray that surrounded him. Worth focusing on for my analysis, I would mention three important "lineages" within this project that converge in not easily anticipated fashion: television, underground film (particularly the cinema of Andy Warhol), and popular gay club music. Thus, there are many members of this family, and avant-garde and popular queer cultures appears with the familiarity of his own relatives throughout the film.

While it met with strong endorsement from many professional critics, *Tarnation*'s reception among viewers who have responded on web-facilitated discussion boards has been divided, with some writers defending the film, while the bulk finding it too "self-centered" and failing to offer any strong narrative. Among the many complaints to be found about *Tarnation* among the respondents on electronically-facilitated viewer's fora is the persistent rejection of the film that can be paraphrased as something like "I thought I was renting a documentary about the shattering effects of ill health and institutional abuse of families but all I got was a lousy music video about growing up queer." This rejection stems, its seems, from the sense of impossibility that a queer artist could educe any value—cinematic, financial, biographical—from the scraps of recorded media forms, much less make a movie that challenges the generic terms by which we might comfortably make sense of the film.

Tarnation is a challenging film, but part of the crisis being described here emerges from the problem of understanding the film as a documentary—that Caouette's relation to his pro-filmic material seems to produce an excess of affect and little "factual" material with which to form any conclusions. Confronted with Caouette's layering of the film, video and acoustic tape that he assembles and shapes through iMovie formatting, many viewers decide that the project is "narcissistic" and not worth giving their time and a place in their Netflix queue—a repeated motif in viewer responses is that the film-maker is a "drama queen." Caouette's decisions to educe his home archive indicates that the repositioning of such media brings with it strong rejections about what kinds of recorded culture seem capable of earning—or not—our own eductive attentions.

I am not sure that I can do much to alleviate such spectatorial dissatisfactions, but I do think that we can learn from these complaints inasmuch as the taxonomical problem—that is, the problem of calling this film a documentary—might more productively be rethought in terms of how *Tarnation* is part of the longer history of another cultural form, that of the melodrama. Here we can see how the logic of eduction, and the concomitant blurring of the consumer's role into that of the producer through prosumption technologies, is not seamless but, like the Stone Building's relation to architecture, disrupts the canonical archive of film history. We see this in the accusations of emotional excess and self-indulgence because these complaints inadvertently find a place for *Tarnation* within the more long-standing archive of cinema by reinscribing its apparent excesses within the terms of the melodrama. With its emphases on family life (itself sometimes nearly claustrophic), its reliance on the sense that identity and experience are most powerfully shaped by the relation between mother and child, its insistence on "over the top" performance as a response to the contradictions of gender and domesticity at hand, and its sustained use of music to organize and underscore (literally) its most important narrative junctures, *Tarnation* operates as a kind of non-fiction melodrama. As non-fiction melodrama, we witness again the strong imbrications of affect and value which materialist feminist critics have remarked upon as a hallmark of the melodrama from its earliest moments in live performance and cinema. Pam Cook, Mary Anne Doane and Linda Williams have theorized how the rewards of maternal labor are the ostensible ability to dwell within the very contradictions of domestic labor: hugely exhausting and barely rewarded in monetary terms, the exchange around women's bodies as childbearers, caregivers and household workers is based less upon a financial transaction and more in terms of the alleged affective returns given to women.

In these terms, what critics of *Tarnation* seem to be emphasizing is its failed emotional economy, to the degree that Renee's distress and ill health are "perversely" compensated by Jonathan's "outlandish" behavior. When we see video footage that Caouette includes of himself performing various fictional female characters that as a child he invented for himself, the film indicates that it is the queer son, and not the mother, who "acts out" in inappropriate ways, ways more associated with drag and the forms of performativity associated with day-time television. Here, Caouette gleans a wrenching performance that signals simultaneously his profound trauma and his already fertile eductive capacities for reading popular culture.

In a sense, rejections of Caouette's alleged narcissism seem to suggest that, whatever forms of aberrant behavior we might witness from Renee (such as when she sings to a pumpkin) these are acceptable as ostensibly recognizable symptoms of hysterical female behavior—what's *not* acceptable is Jonathan's embrace of those excesses in his adolescent video enactments. Claims about *Tarnation* as being self-indulgent, as the work of a "drama queen," then might better be understood in the film's refusal to frame the gender-play at work in these early video performances as part of a pathology; one can imagine a more recognizably "straight-forward" narrative that would seek to understand how a queer child might become a queer adult and lay blame upon the mother, such as that found in Alain Berliner's 1997 French language feature, *Ma Vie En Rose*. Rather, in a caveat to regulatory panics about the fears of mass culture on the developmental lives of children, Caouette offers the powerful insight that television might, helpfully, make you into a drama queen when nothing else is going your way. In a sequence about the rich fantasy sponsored by his adolescent attentions to cinema and tv, the *other* family members who raised him about whom I spoke a moment ago—the family of the archive—collide in a montage of Hollywood film, made-for-tv-movies, children's tv programming, 80s music videos and Warhol cinema.

Tarnation does not, however, solely reveal the eductive possibilities of the archive for the filmmaker, but expands its attributions of value through the networks in which the spectator is situated as well, and important to emphasize are those aspects of the film that shape its eductive dimensions for us, its viewers. First, I would draw attention to the relation of the director's commentary to the film proper, for in this regard *Tarnation* is, like so many of its digital cinematic kin, not one film but many. The accompanying "extra features" of its DVD release offer a palimpsest of its presumptive activities when we discover Caouette's comments about how he achieved certain visual and acoustic effects using iMovie, or the fact that he found some of the audiotapes that he uses behind his grandmother's clothes dryer. Where to find potentially eductive materials in our own mediascapes (i.e. look behind your dryer), and the possibilities for manipulating such forms on our personal computers situates *Tarnation* as a pedagogic text that allows us to queer our own archives. Second, Caouette also tells us that he couldn't secure the rights to particular popular music with which to score the film, and by indicating those songs he might have preferred, the film makes it possible for the viewer to "rescore" the film by locating the pieces he might have used by downloading them to listen to alongside the image-track of the "official" version. Indeed, an examination of playlists compiled by some viewers and posted on blogs reveals the manner in which the eductive economy of this archive already unfolds through other media networks and through the efforts of participants other than Caouette. Put another way, once the process of eduction begins, it often enlists the efforts of others working in various media forms—iPod users making playlists, fan-composed blogs, indeed, even mental health workers who, as in the case of *Tarnation*, have been worried about its use as a self diagnosis tool by any potential viewers and have discussed such on health-related list-serves.

By way of moving to a conclusion, I would historicize the effects of presumption and value's eductions by considering *Tarnation*'s archive in relation to that of more large-scale studio film. It is remarkable that the corporate entities that produced such forms historically almost never saw any occasion in which to organize their productions for redistribution; Hollywood in the classical period obeyed the corollary of terminal consumption by maintaining a near indifference to its back-catalog until quite recently. The industry's shift towards an understanding of the cultural commodity as a site of possible eductions occurs, I would suggest, because of what digitalization makes possible: the dispersal of the archive back into the home, the place where, as Derrida notes, it was originally housed in Hellenistic societies. Read in terms of the modes of exchange at hand, the archive's return to the domestic sphere also makes the "economic" palpable in its etymological roots—the home, or "oikos" in Greek, becomes the renewed scene for activities of manufacture and decisions about labor. The possibilities for such dispersions into the domestic meets their own dialectical turn in the dual tendencies of the industrial sector both to take advantage of the economies of scale to minimize production costs and maximize profit by taking advantage of the digital's iterations while simultaneously attempting to enforce the sense of the commodity's terminus in the sphere of consumption, a terminus that, for the cultural commodity, now seems not to have existed in the first place.

Thus, film's digital incarnation behaves in a new and historically unpredictable way: it reveals the possibility of cinema's circulating at different moments in its movement through the economy as only occasionally a commodity. Eduction involves the extraction of new forms of value, sometimes monetary and sometimes not, from the archive, and allows for the fact that monetary value itself is increasingly a mediated category. More directly to the point of the digital film that now makes its way into the domestic archive, say in the form of the DVD, the film industry relies upon the home user to participate in the new extraction of value heralded by eduction. This may occur through a variety of motivations, such as nostalgia, historical curiosity, even boredom, but some form of intellectual effort is necessitated by the foregoing form of cinema in its economic and intellectual dispersal via the digital. Thus the digital object has unleashed a new kind of exchange, based frequently less on decisions about money than on time, and such exchange is made possible by the very characteristics that make digital forms so easily circulated in the first place. It also compels that figure who might formerly have been called the consumer now to work upon the text. This compulsion occurs in a number of ways— through fan commentary on message boards, in the distribution of image and sound files on personal websites, in the production of non-corporate cinema, in the unlicensed reworking of the corporate film product, in the choices that the dvd user makes about aspect-ratios, deleted scenes and director's commentaries and, finally for my topic today, in the non-industrial film-maker's assemblage of the personal archive.

Put another way, the digital object of the moving image or the song file "stores" value—not necessarily in a monetized form—and increasingly seems to place no demand upon the economy to be adjudicated through the nexus of monetary exchange. However, in its capacity to store such value, other measures must come into play in order to understand relative worth, and it is the activities of eduction that devise those measures. At this point, a few disclaimers are in order: first, the category of eduction indicates those efforts in which value is extracted from the materially abiding digital commodity, and does not supercede the categories of production and consumption as much as it is meant to describe and theorize a particular form of value-coding at work when those two spheres collide in the digital era. Second, this is by no means a utopian process, as eduction gestures towards several important contradictions at hand: that the eductive subject will probably not realize any monetary value from his or her work, and simultaneously, the corporate sector is busily devising its own strategies for this new archive in the hopes that it can sustain a sense of centralized proprietorship. Lastly, digital eduction does not necessarily imply any concomitant progressive or radical politics, but it does make room for those cultural producers situated as subjects of alterity to gain access to the archive in new and unheralded ways.

The process of renovating Stone's 2 Columbus Circle is nearly complete, turning it into what one critic describes as "New York recast in the image of an office park for Swiss pharmaceutical companies." The building must now become part of a different archive, the one we carry around in our heads of the things we could not preserve—a virtual archive. The loss of one building in mid-town Manhattan may not be the most egregious thing to be visited on New York, despite the fact that, as Simone Signoret once said in another context, "nostalgia isn't what it used to be." Yet, framed in the terms offered here, it might make us pause, especially those of us who work on a far more vulnerable and, I would say, intimate archive, that of the moving image. Yet, I would also argue that, in a world where the mess of a mentally unbalanced family can be turned into a widely-distributed film by the queer eye of a queer audodidact, where the film industry's accountants shudder at the prosumptive future and its implications for its own survival, it serves us all well to pause to rethink our own field and its underlying issues— spoken and assumed--of "value" and "archive." What we address as scholars and teachers is the everyday life of disposable ringtones, websites, blogs, chat rooms, mashups—in other words, the potential Blockbusters, Classics, Masterpieces, and Academy Award winners of our mutually constituted future and its eductive possibilities.

Endnotes

1 Patricia Zimmerman, *Reel Families: A Social History of Amateur Film*, (Bloomington: Indiana University Press, 1995) xv.

2 Jane Simon, "Recycling Home Movies," *Continuum: Journal of Media and Cultural Studies*, Vol 20, No. 2, June 2006, 195.

3 More to the point, one of the driving tenets of *Moving History* was that materials "be available for use online without copyright fees or other access restrictions." Frank Gray and Elaine Sheppard, "Moving History: Promoting Moving Image Archive Collections in an Emerging Digital Age," *The Moving Image*, vol. 4, no. 2, pp. 114.

4 http://www.movinghistory.ac.uk/archives/sw/films/sw6fashion.html

5 http://www.movinghistory.ac.uk/archives/sw/films/sw1dresses.html

6 Herbert Muschamp, "The Secret History," *The New York Times*, January 8, 2006, p. 35.

7 Jacques Derrida, *Archive Fever: A Freudian Impression*, translated by Eric Prenowitz (Chicago: University of Chicago Press, 1996) 2.

8 A longer version of this paper historicizes eduction in relation to other theoretical considerations of value-coding.

Theorizing Interface

Metaphors help us comprehend how digitality weaves, binds, encloses, bridges, spans, and navigates across technologies, spaces, and disciplines (from genomics to urban planning).

Sylvia Nagl and Sally Jane Norman
"Raranga Tangata: The Weaving Together of People"

Nicole Starosielski
"Reskinning the Digital Surface: Borders and Immobility at the Interface"

Sarah Sweeney
"Way-Finding on the Web: Urban Planning and the Virtual Interface"

Raranga Tangata: The Weaving Together of People[1]

Sylvia Nagl and Sally Jane Norman

Collaboration favoured by twenty-first century information and communications tools is still largely subservient to and inhibited by behavioural patterns carried over from last century. Entrenched specialist enclaves remain deaf to the multiple voices and translational dynamics resonating at interdisciplinary crossroads. Jealously maintained territorial walls remain blind to the cognitive windows opened up by new kinds of exchange. Shifting bodies of collectively shaped, constantly emerging and evolving knowledge loom like uneasy shadows over those who stubbornly wield bygone forms of authority as exclusive and unchallengeable. To speed us beyond such inertia, we need to create inspiring models of encounter that are tuned to the sociality offered by today's technologies. These models must foreground rather than merely tolerate polyphony, difference, ambivalence, and contradiction, in order to build fittingly humanised information agoras.

We propose an experimental model which aims to explore the rich diversity of mappings and readings that surround embodiment. As paired cross-disciplinary presenters, our starting point is at least twofold: genetics and bioinformatics is one of our main strands, art and creative visions of the body is another. Yet these specialisations are in turn woven into willfully interdisciplinary fabrics of thought and a shared sense of urgency to develop singular forms of embodied knowledge.

Raranga Tangata: the weaving together of people. This Polynesian expression, used to designate the Internet, is one of many powerful poetic testimonies to the living culture of the Maori people of Aotearoa – New Zealand, a culture deeply meaningful to both presenters. Polynesian cosmogony vividly shows how a collectively shaped and transmitted narrative can offer cognitive handles to those seeking meaning amidst the chaos of complex worlds. The Maori creation myth revolves around the concept of "whakapapa", or genealogical layering, to expound the series of events whereby humans first emerged, whereby the first bodies were born and made through three states of evolution: Te Kore; energy, potential, the void, nothingness; Te Po; form, the dark, the night; Te Ao-marama; emergence, light and reality, dwelling place of humans. Polynesian culture is deeply embodied and anthropomorphised, from its narratives of primal surroundings to those that describe human development and evolution. It offers viscerally recognisable readings of complex processes, through the creation story from Te Kore to stories of kinship (iwi, hapu and whanau) then individuals. Pūrākau (mythological traditions) are statements about the nature of the world which echo the creation story, so that the world is ritually 'recreated' whenever creation whakapapa (genealogies) and kōrero (stories) are recounted. The Maori stand amongst the world's finest navigators, and their mapping and steering skills are as marvellously reflected in the meaning-making weave of their stories, as in their path-finding journeys across the Pacific Ocean.

In contrast with the thousands of years of cosmological and physical mapping that are hallmarks of Maori culture, complex systems of a new kind have been the object of a steadily growing field of research over the past decades. Complexity unites the grand challenges humanity faces at the beginning of this new century. From climate change to food security, the global economy, global politics and conflict resolution, ICT networks reaching across the planet, emerging epidemics and health - these examples span hugely disparate scales, but all of them are manifestations of complex systems, and the enormity of the challenges is unprecedented in human history. The cognitive resources and investigative practices which have successfully informed human agency in the past are greatly unequal to the realities of the 21st century. This problem is exacerbated by the persistence of local knowledge systems insulated from each other to a greater or lesser extent - for example, science, humanities, arts, technology, as well as, very importantly, knowledge held in different cultures.

Consequently, what is urgently needed is massively intensified exchange and integration across all disciplines and across cultures with diverse worldviews and richly diverse cognitive, material and social resources for addressing the challenges arising from our embeddedness within complex systems and our own embodied nature as complex systems. A paradigm of 'complexity' is paving the way for narratives which integrate concepts and metaphors including system, holism, inter-connectedness, multiplicity, interaction, network, dynamic change and emergence.

Emergence is a particularly potent concept as it opens up alternative, and potentially revolutionary, perspectives on embeddedness and embodiment. It defies traditional epistemologies of causality, assertions of single causes and privileged loci of control, including any assumed primacy of the genome as a blueprint or a program. Emergence re-focuses our gaze from the fragmented body to the whole, from the reduced and uni-dimensional to distributed, complex, local-global unity; emergence in the body seems 'machine-like' and 'organic' at once. Like the creation whakapapa, it evokes the coming-into-being of a coherent, self-organising, self-sustaining system with complex structure and behaviours, thanks to multiple, parallel interactions between entities. In the dimension of space, an emergent system is seen as made up of hierarchical layers of increasing complexity, from molecules to cells, organs and the body, and in the dimension of time, it undergoes state changes at local and global scales.

So how might advanced mathematical models and computer simulated processes of emergence be wrought into meaningful visions spanning the sciences, the humanities and the arts? How might multimodal and immersive technologies enhance cognitive fluidity and enable engagement with intellectual, cultural and artistic complexity in thinkable, tangible, visualisable ways?

Complexity/emergence interfaces with evolution, development, technologies of information and the human genome, mythological creation stories, artistic and cultural readings of embodiment. These interfaces can be sealed or permeable, they can be fault lines of tension and struggle or places of exchange and shared creativity, they can offer openings for exploration of a rich diversity of mappings and readings that relate to embodiment. Assertively poetic, productively ambivalent narratives can inspire us to explore our newly created electronic territories of collaborative social encounter. Navigational tools creatively fleshed out with embodied knowledge to prioritise sensory and experiential integrity in these times of discretely disincarnated media may provide invaluably effective and affective inroads into our info-rich world. Artistic endeavour fundamentally addresses the need for diversified worldviews and materials, since art uniquely enfolds multiple layers and sometimes fertilely contradictory voices, lending itself to and building upon difference. Like mythological systems, art works are openly interpretable and uniquely holistic in their crafting of poetic experience, yielding readily grasped idiosyncratic perspectives.

We propose risking a moment of uninhibited creative conjecture, an attempt to flesh out an interdisciplinary story of embodiment drawing on two strands of thinking: genomics and complexity science, and artistic narratives. These strands weave a poetic narrative, a fabric to grace the shimmering, changeling contours of our electronic techtonic world. Raranga tangata.

Endnotes

[1] The authors are grateful to Charlie Tawhiao for having communicated and defined this term: « I prefer the metaphor approach, so I consider a network of people such as that presented by the internet to be a weaving together of people similar to how a mat is woven: *raranga* or *whiriwhiri* refers to the weaving of a *whariki* (mat) or *kete* (basket). The internet community could therefore be described as *raranga tangata* or similar to describe the weaving together of people.» Personal correspondence, CT – SJN.

Reskinning the Digital Surface: Borders and Immobility at the Interface

Nicole Starosielski

I want to pause, here, at the skin of the interface. What does it mean that skin is the metaphor through which we encounter digital surfaces? How does this term, "skin," affect the way we make sense of the interface? And just as our view of the interface is affected by the language we use to describe it, our understanding of skin itself is discursively constructed. Its meaning fluctuates with historical and cultural context. It might be layered or homogenous, a barrier or a penetrated space, a reflection of, a stand-in for, or the extension from this interior. The way skin is depicted, represented, and made sense of reflects specific relationships of the self to the world. Thus we can ask: which kind of skin is the skin of the interface and, through this question, explore what kind of relationship between the technological body, its surface, and the user's body lies implicit in contemporary discourses.

In technical discussions, popular texts, and many new media theories, the dominant tendency has been to treat the interface skin as less important, interchangeable, and insignificant in light of the system. If the skin is marked as significant at all, it is typically for its function as a permeable and porous membrane, and is as far as it facilitates access to this "authentic" interior. This understanding of skin is valuable in that it counteracts a longer cultural tradition which has increasingly portrayed skin as a rigid border, and the emphasis on skin's transgression enables the delineation of new types of subjectivities, especially from a feminist perspective. However, understanding the skin as simply permeable becomes problematic when we look at the type of space the digital skin tends to represent: it is a transparent space, existing only in and through the user's penetrations. In addition, these skin discourses distinguish the digital ontologically from other media skins and surfaces, which becomes difficult as digital media is increasingly embedded in the environment and convergent with other media. I would like to suggest that an attention to the digital skin as a selective border is both an important part of a feminist critique, as it helps to recuperate immobile experiences at the interface, and can help break down the boundary between new media interfaces and the surfaces of other media, such as film and video, and the divide between theories which address them.

In her book, *Skin: On the Cultural Border Between Self and the World*, Claudia Benthien traces various discursive shifts in the way skin has been understood as a symbolic surface between the self and the world. Paradoxically, while medicine has exposed the interior of the body, skin has become understood as an increasingly rigid boundary, and in the twentieth century, the "central metaphor of separateness."[1] She identifies two primary levels of the articulation of skin's meaning, each which points to diverging conceptions of subject and body. On one hand, the skin is an outer shell, a sheltering, concealing, or deceptive cover which is other than, and foreign to, the authentic self underneath. The second conception of skin equates the skin with the subject; skin metonymically stands in for the whole being. The expressions and language of this second conception, where skin stands in for self, has diminished over time in favor of skin as separate, deceptive, and rigid boundary.

In discourses of computer science and design, skin is most often configured as an insignificant layer over the "authentic" system. PROSKIN, a research project optimizing a skinning tool, defines the skin as follows:

> A skin is considered to be the appearance of the user interface, including graphic, haptic, and/or aural patterns…Skins are used typically to change the "look and feel" of the interface components, often a cosmetic change alone (i.e. the colours change or a background image is applied, but the interface components remain unaffected in location, attribute or function.)[2]

This layer, though distinct from the system, is far from the rigid boundary Benthien describes. It doesn't separate the system from the user's body, but rather, is an interchangeable, superficial, or cosmetic element without any effect on the relationship of interior to exterior.

This perspective is mirrored in popular discourses. A reviewer in the *New York Times* writes, "Skins are faceplates that cover your MP3 player like masks, creating a visual appearance of your choice. There are hundreds to choose from, and you can switch between them as often as you want."[3] Skin is not significant in itself, does not affect the user's experience of the new media object. Unsanity.com advertises their skinning program: "You don't wear the same clothes every day, your house doesn't look exactly like your neighbor's - why should the computing interface you use every day be any different?"[4] Skins, just like our own, are recognized and defined by difference. However, the skin's difference mirrors the user rather than the computer. Mark Rolston, the VP of Frogdesign writes, "You sit in front of (your computer) all day … and it represents you…"[5] Thus, on one hand, "skin" describes the part of the interface that is replaceable and disconnected from the authenticity of the system, it also becomes a reflection of the user. In this way, skin is not considered as important to phenomenological experience, but rather, is discursively understood as a space to be colonized by the user.

This tendency to downplay the role of skin might be attributed to the structure of programming, as the interface's look and feel is easier to alter than system functionality. However, it can also be understood in the context of, and enabled by, cyber-structuralist thinking that has disregarded the importance of borders and surfaces. In "the annihilation of time/space discourse" characteristic of early new media theory, Marshall McLuhan and others hailed the potential of new media to overcome temporal and geographic boundaries, as well as the material limitations of their own bodies.[6] McLuhan famously writes, "[i]n the electric age we wear all mankind as our skin."[7] In these discourses, the focus is the system and its ability to propel the subject. Like the senses I outlined before, the actual skin of the interface is only significant to the user experience as a means to extend their own reach: both the interface skin and the skin of the subject is transcended.

This emphasis on transcending the skin is echoed in formulations of the cyborg and the post-human. Donna Haraway writes, "[h]igh-tech culture challenges these dualisms [of human and machine] in intriguing ways. It is not clear who makes and who is made in the relation between human and machine…Why should our bodies end at the skin, or include at best other beings encapsulated by skin?"[8] The skin, both that of the interface and that of the subject, is as a boundary to be trespassed. Claudia Benthien argues that this rhetoric of new media is characterized by the breaking down of the previously rigid formulation of skin as border. Skin becomes trespassable and broadly penetrated.

Recent new media phenomenologies have reinvested the material surface of the media object and the material body of the user. As one example, Anna Everett, in her theory of digitextuality, argues that the "click fetish" of new media lures the body with a promise of sensory plentitude. It is the user's click at the site of the interface that draws them into the hyperlinked space. For Everett, and a number of others, the surface of the interface is significant as the space in which bodily and sensory experience is generated: the skin is the concrete place in which the body's moves are made. Interface skin is important because of its materiality: rather than being simply transcended, it is understood as permeable, porous, a means of input, output, exchange and mobility. It is the presence of this skin which makes information accessible, and which allows us to understand ourselves as penetrating the system in the first place.

On the whole, then, we can extract two levels of meaning from this range of discursive examples. First, the skin understood as interchangeable and replicable, a space for the user to insert herself. And second, the skin is constructed as a permeable, porous, membrane, the space that enables the user's movements. As Benthien recognizes, this second mode is a valuable corrective in light of breaking down the rigid boundary between self and other, between bodies, and between bodies and digital media objects.

This mode of understanding the skin of the interface, its appearance, can be problematic if taken by itself. In a longer version of this paper, I use Gillian Rose's feminist work on geography to articulate how these discourses understand the space of the interface (and bodily skin). In geographical terms the permeated interface skin is a transparent space which allows action and energy to pass through. It is a space to be traversed, and conceptualized in the movements through it. Skin, however, does not exist only in its connections, and simply thinking about how digital media skins enable us to move does not address experience which is isolated or immobile. While these discourses frequently deem irrelevant, or interchangeable, or non-affective, the parts of an interface skin which do not serve a function in connecting or transporting the user, I would like to argue that it is what skin keeps out, what it refuses movement and "immobilizes" that in some sense defines it. An attention to permeability must be complemented with an understanding of skin's selectivity, the way in which the interface is a perceived boundary which, at points, we cannot trespass. It is this meaning, skin as a border zone, which is infrequently attributed to interface skin by popular media, but more often addressed by hybrid new media artwork.

I want to bring up here, as a counterpoint, the way in which the video surface has been formulated as a "skin." In her book *The Skin of the Film*, Laura Marks theorizes the surface of film and video as a skin partially through an exploration of "haptic images." In contrast with optical images, which represent a three dimensional, symbolic place that the Cartesian viewer imagines as extending their space, "haptic images" are incomplete and partial, fragmented or blurred. Rather than plunging into the depths of the diegetic world, our gaze rests on the skin of the screen, distinguishing its textures and patterns. Marks suggests that viewing haptic images may be more like a mode of touch, evoking our other senses, and our bodies. The viewer is called on to fill in for the image's gaps, engaging with its traces. Thus, an "immobile," but fully embodied viewer, is drawn into an affective relationship with the "skin of the film." The viewer comes to understand the media surface as a skin, as another body, precisely because he or she is not allowed through it.

Here, I want to insert an object for discussion which problematizes the dominant understanding of digital skin as penetrable, easily transcended, and defined in opposition to the less permeable or impenetrable skin of other media forms. The video game *Playas: Homeland Mirage*, foregrounds the way in which the skin itself, its appearance and its function as a selective border can immobilize the user. The game play is set in the real-world desert town of Playas, New Mexico, which was purchased and converted to an anti-terrorism training facility. Now host to a variety of U.S. military simulations, the few residents left over have a choice to either role-play in the simulation or stand by as onlookers.[9] The game itself takes place on a street of the town populated with terrorists, civilians, and "Department of Homeland Insecurity" agents. The game play itself is limited and circular. There are no levels. There are no goals. There are no significant actions that the player can take. They can only haphazardly activate video clips, fragments left over from the residents' lives. Thus, the player's experience is characterized by an alternation between on one hand, agency and movement, and on the other, being gunned down. We can think of this metaphorically as an alternation between being allowed to selectively trespass and being held up at a border. The game thus schematizes the different functions of the digital media skin: permeability and selectivity. These modes of engagement are meant to mirror the situation of the real world Playas, where the player can either take the role of the spectator or the victim.

The physical construction of the game mirrors the experience of the game world. The game is projected onto a screen in an enclosed room, and while only one player is allowed to navigate the world, it is built for an audience. Sensors register the other people in the room and project a blurred reflection of their movements onto the screen. The audience members can only trace the remnants of their reflection over the depths of the game world. Within the physical game space, there is thus a play between movement forwards, in and through the room itself, and an attention to immobility, a captivation at the surface of the screen.

The way in which this immobility is made obvious to the player and the spectators is through a use of haptic imagery and a redirection to the surface itself in the way that Laura Marks describes the viewer's relationship to the skin of the film. On one level, the longer the player engages, the more the game environment fragments and blurs. Both this mirage aesthetic and the mirroring of the audience members render an incomplete, impressionistic and sensory world. The appeal is not so much in understanding the meaning of the characters' movements and the three dimensional space of Playas, as it is watching the textures of the interface, the unexpected blend of user-reflections and the distortion of the landscape. A major component of the experience is also the haphazard activation of video clips, themselves distorted remnants of the residents' lives. They are not the typical cut scenes of the video game, however, where the game play is stopped and the narration begins, but rather are mapped onto the surfaces of the world and can be left at anytime. While on one hand, they appear to extend our space, on the other they force us to call upon our own bodies, histories, and senses to fill in the traces left by the residents of Playas.

While we approach *Playas* with the expectation of navigating into a world, searching out targets, and fulfilling objectives, these desires are only partially fulfilled. At times, we are immobilized at the surface of the screen, in the space of mirage, distortion, and haptic imagery. We cannot simply search-and-scan for the relevant information, but rather are confronted with the limits of immersion, interaction and knowledge. *Playas*'s implicit critique of representation, and the penetration of the skin supports an explicit political critique of the circularity and lack of information flow in our contemporary political situation, as well as American penetration into other geo-political spaces.

Thus, in *Playas* the digital skin is configured as a selective border, as a borderland, where the viewer resides, unable to penetrate into space. This immobility is productive, precisely because it forces us to call upon our own bodies to fill in the gaps. This discursive example foregrounds its own surface, not as an interchangeable graphic pattern, but as a significant space which sometimes rejects our attempts to enter. I thus want to argue for a theorizing of the interface skin as a border, and for the significance of the skin as an affective, and potentially embodying, layer of the interface. Not just a space to be colonized and penetrated, but a possible space of resistance.

I want to conclude by noting that this is not simply a rhetorical move. Just as this metaphor of skin is used to understand the interface, we increasingly use the metaphors of the digital media interface to make sense of our own skin. Claudia Benthien writes, "[t]he epidermis, the largest human organ in terms of surface area, is being discovered as an interface."[10] This presentation, and hopefully this discussion, is a step towards a critical fleshing out of their interconnections, a move to articulate how contemporary perceptions of the border between technology and ourselves, our interfaces and our skins, are discursively intertwined.

Endnotes

[1] Benthien, Claudia. *Skin: On the Cultural Border Between Self and the World.* Trans. Thomas Dunlap. New York: Columbia University Press, 2002. p. 1.

[2] Fine, Nick and Willem-Paul Brinkman. "Avoiding Average: Recording Interaction Data to Design for Specific User Groups." *Entertainment Computing - Icec 2004: Proceedings of the Third International Conference.* Berlin: Springer, 2004. p. 399.

[3] Strauss, Neil. "The MP3 Revolution: Getting With It." *New York Times*, July 18, 1999.

[4] Unsanity's website: http://www.unsanity.com/haxies/shapeshifter, Last accessed April 5, 2007.

[5] Quoted in: Kahney, Leander. "How Mac OS X Can Shed Its Skin." *Wired Magazine*, December 12, 2003.

[6] I borrow this phrase from Lisa Parks, "Kinetic Screens: Epistemologies of movement at the interface." *Mediaspace: Place, Scale and Culture in a Media Age.* eds. Nick Couldry and Anna McCarthy. London: Routledge, 2004.

[7] McLuhan, Marshall. *Understanding Media: The Extensions of Man.* Cambridge: The MIT Press, 1994. p. 47.

[8] Haraway, Donna. "A Cyborg Manifesto: Science, Technology, and the Socialist-Feminism in the Late 20th Century." *The New Media Reader.* eds. Noah Wardrip-Fruin and Nick Montfort. Cambridge, MA: The MIT Press, 2003. p. 532-33

[9] Hall, Mimi. "War on terror takes over a thankful town." *USA TODAY*. 3/13/2005.

[10] Benthien, Claudia. *Skin: On the Cultural Border Between Self and the World.* Trans. Thomas Dunlap. New York: Columbia University Press, 2002. p. 6.

Way-Finding on the Web: Urban Planning and the Virtual Interface

Sarah Sweeney

When I open a new browser window I experience a simultaneous surge of excitement and dread. In the blank screen and blinking cursor there is the possibility of accessing billions of sites, but there is also the frustration that I do not know how to find most of them. I began this project thinking that I was alone in this frustration since there are so many critical and popular works extolling the virtues of our new personalized, more searchable, ever-changing web. However, as I kept looking, I found statistics and anecdotal evidence that suggested that there are many of us, users who are overwhelmed by the options and discouraged by our current tools for navigating among them. In this presentation I will speak as a web user and as an artist, nothing more. Kevin Lynch's seminal urban planning book, *The Image of the City*, will provide the framework for much of my argument. It also serves as a guide for my tone, an approach he describes as "speculative and perhaps a little irresponsible: at once tentative and presumptuous."[1]

In this talk I will take interface in a broad and malleable sense, as standing for the different technological elements that provide a connection between users and the websites they visit. I do not want to focus on the navigational elements that we use once we have made the connection to a website since this area has already been extensively covered within studies of usability. Instead, I would like to focus on the familiar yet commonly overlooked interfaces that we use to navigate our way to and between sites. These interfaces are difficult to define and identify since they can take many forms, including lists of links and form elements. They also have different levels of immersion within the web environment—some exist as independent entities like search engine widgets or the address bar while others are embedded within websites and even within web content. While these interfaces differ in location and form, their common goal is the element that unifies them as a group—they all work to help a web user find his or her way to a web site.

It is this same goal of finding one's way that Kevin Lynch studies in his 1960 book, *The Image of the City*. In this text Lynch uses this concept of way-finding as a means of assessing different systems of urban planning. Lynch suggests that "[i]n the process of way-finding, the strategic link is the environmental image, the generalized mental picture of the exterior physical world that is held by an individual" (*IC* 4). This image must have several qualities to have what he considers "value for orientation in the living space." It must be sufficient, clear and well integrated, safe with the possibility of alternative actions, communicable, and adaptable to change (9).

Thinking about the web through Lynch's criteria initially suggests that the interfaces we most commonly use would not provide us with a strong environmental image. I would suggest that Lynch's first quality of sufficiency, which he describes as "allowing the individual to operate within his environment to the extent desired" (9), is perhaps the most essential quality in an operable environmental image and also perhaps the case in which web interfaces fall the most short. In the most obvious and basic case of this operation, finding one's way to a destination, these interfaces generally accomplish the goal in question without relying upon an environmental image. For example, the address bar, search engine interface, or a user-determined portal such as Alexa or Digg will very often take us directly to a variety of different destinations. The problem arises when we would like to return to this destination after some time has passed. Unless we bookmarked the destination or remember its address verbatim, without an environmental image to orient ourselves we must retrace the path we took exactly. However, since the path generated through these types of interfaces is generated at a specific moment in time it changes over time to reflect the changing relations within data on the web. The rankings of search engine results can change from one day to the next and what appears on the first page of Digg or as a top site on Alexa can change even more rapidly, making it difficult to retrace our steps in many cases. While a stable environmental image is not completely necessary for every type of navigation I would suggest that without it we lose a crucial element of control over our environment in some situations.

Just as time can change the way a single interface represents the relations between information, each interface also represents those relations differently based on its own organizational methodology. A search engine interface might rank and organize sites based on a page-ranking algorithm while a social interface like Alexa might organize sites based on a different criterion such as traffic. Thus the same site could appear through each interface in very different contexts. Because of the different ways in which these various interfaces mediate the relations of the web, any given site exists within a multiplicity of contexts, which works against the creation and integration of a clear, operable environmental image.

Without this clarity and integration, the communicability that Lynch also considers a crucial element of a successful environmental image becomes difficult. While it is relatively easy to communicate the address or name of a particular site, it is far more difficult to communicate the path necessary to reach a site without this type of specific information. In the absence of a clear and readable environmental image we cannot refer to common experiential or contextual clues that could help us to direct others when we no longer remember a precise destination. There are specific interfaces designed to help you share addresses without relying upon your memory—for instance a personal bookmark manager or social bookmark managers such as del.icio.us— but they solve this problem without helping to construct an actual environmental image. I would suggest that while aiding our navigation, these interfaces force users into a more narrow and precise form of communication than might be possible with an environmental image.

Another quality Lynch considers important is the safety that comes with having multiple ways to reach a destination. He writes that an environmental image "should be safe, with a surplus of clues so that alternative actions are possible and the risk of failure is not too high" (9). He warns that "[i]f a blinking light is the only sign for a critical turn, a power failure may cause disaster" (9). Given the text-based nature of web navigation, so can a misspelled URL. This text-based nature creates a need for precision of language when we enter a search term or a web address. While Google has built in a corrective mechanism that asks you if you might mean to be searching for one thing rather than another, this is only a single option that is controlled by the technology rather than the user. Perhaps it is largely because of these limited options that, as one study suggests, "only one in ten professionals always finds what he or she is looking for on the first attempt," while almost 70% of the same group admits to "end[ing] up on sites they didn't expect to visit and are not relevant to their work."[2]

The last quality that Lynch cites as crucial to a functional environmental image is adaptability to change. Given the shifting nature of the web through the differences in time and interface I have discussed above, it would seem that our current web interfaces easily meet this criterion. However, I would suggest that this is not the kind of change Lynch is thinking of. While our current interfaces do indeed respond to the changing shape of the web, the changing elements of the image are often more numerous than the stable elements. For example, a search engine rearranges the results of the search with each new request. This constant rearrangement of our environment does not create an open-ended environmental image that is adaptable to change. Instead it makes the development and retention of an environmental image difficult if not impossible.

From this application of Lynch's theories two important conclusions start to emerge. First it becomes clear that the web interfaces that we commonly use make it difficult for a web user to produce a functional environmental image. For Lynch this would mean that while it is still possible to navigate this type of space, without an adequate environmental image we can only do so with what he describes as "the cost of some effort and uncertainty" and in even more extreme terms as strain, anxiety and even the terror that comes with complete disorientation (IC 5, 4). However, the second conclusion we reach is that these same interfaces accomplish many of Lynch's goals for an environmental image without actually producing one or causing the strain or anxiety Lynch predicts. It would seem at first glance that these two conclusions suggest that Lynch's mid-century theories of navigation within a physical space are inapplicable to the virtual space of the contemporary web.

However, I would suggest that the problem is not that Lynch's theories are not compatible with our experience of way-finding on the web, but rather that his theories do not take into account the way in which different processes of way-finding are influenced by different variables. Crucial among these variables is the degree to which we can articulate our destination. In 1960 the only possible process for way-finding was to navigate through one's environment from origin to destination regardless of the certainty with which one could identify that destination. Lynch's work could not anticipate a process of way-finding that allows for a direct connection between origin and destination made possible by the collapse of space within the virtual environment of the web at the end of the century. Even the fiction of instant connection to a destination popularized by *Star Trek* in the form of teleportation was six years away.

Thus while our ability to articulate a destination was not a central issue for Lynch in 1960, it is central today, as it determines how we choose between the different processes of way-finding. The degree to which we can articulate our preferences and its role in our decision-making process is discussed by Alexander Chernev in a 2003 study of product assortment. In this study Chernev finds that "individuals with a salient ideal point face the relatively simple task of searching for the alternative that best matches their already articulated attribute preferences."[3] I would suggest that we find the same processes at play on the web. In a situation where we can articulate the ideal destination, getting there is simply a process of finding a match. The existing interfaces accomplish this goal easily without reliance upon an environmental map. Moreover, to use an environmental map in this type of situation would only complicate the process and make it less efficient.

However Chernev also finds that "individuals without an articulated ideal point face the more complex task of evaluating the available alternatives while at the same time forming the very criteria to be used in the evaluation process" ("PA" 159). Similarly, on the web the task of finding a destination that is not highly articulated is far more difficult than finding an articulated one. In such a situation both our alternatives and the criteria derived from them are generated entirely by our interfaces. In this scenario we lose the agency we had in the case of the matching scenario. Here our destination is predetermined by the alternatives suggested by our interface using its unique organizational methodology. I would suggest that in this situation where our destination is not highly articulated a navigational interface that builds a stronger environmental image might allow us to regain some of this agency by allowing us to define our own alternatives and thus the criteria by which we determine this destination.

In addition to providing criteria by which to judge the strength of the environmental image, Lynch also outlines certain elements that the organizer of an environment, such as an urban planner or interface designer, could emphasize to produce a stronger and more legible environmental image, such as paths, nodes, edges, districts, and landmarks. I would now like to look at some of the web interfaces, real, imagined, and proposed, that employ these elements.

One of these elements is the district, which Lynch describes as a space that is "recognizable as having some common, identifying character" (*IC* 47). While there is no exact online equivalent, a broad application of this concept could include interfaces such as blogs in which links are arranged around an identity, usually an identity constructed by the producer of the site. The lists of sites included within such an interface form a sort of virtual district made distinct and recognizable by the identity that brought them together. The legibility and accessibility of this identity makes our choice between districts more deliberate and informed, thus giving us greater control over our environment.

Another element that adds legibility is the landmark. Lynch notes that the use of the landmark "involves the singling out of one element from a host of possibilities" (48). Two interfaces that differentiate between sites in this manner are Alexa and Digg. Both of these interfaces use the web community to identify sites that take on a different scale in the landscape of the web. While the temporal nature of this landscape makes it difficult to use for orientation over a span of time, at any given moment they allow us to create a clear environmental image that is common and shared.

While both web districts and landmarks provide contextual information that makes it possible to see distinct sets of data or pieces of the environmental image, without a spatial interface it is difficult to integrate them and see the connections between them. A spatialized environment would provide an opportunity for the paths, edges and nodes that create the continuity between pieces, making them more than a series of isolated entities. There is currently no interface that completely fulfills this possibility. However there are several promising interfaces that begin to suggest the potential of such a perspective.

One such space is the fictional world of the Matrix described by William Gibson in *Neuromancer*. This world is spatially organized around a "graphic representation of data abstracted from the banks of every computer in the human system." It appears as "lines of light ranged in the nonspace of the mind, clusters and constellations of data."[4] Another such space is Second Life, described on its homepage as a "digital world imagined, created, and owned by its residents."[5] This three-dimensionally rendered world is structured around a geographical space that has been purchased and developed by the residents. Google Earth is another example of a spatialized interface, although it is organized around a much more familiar map of the actual physical world.

To this group I would also add an experimental interface named ColorColony that I have been developing with my colleague Danielle Laplante. Although we began work on ColorColony before I had read Lynch's work and thus it is not a direct response to his theories, it is a response to the same problems and issues Lynch documents in the least imageable urban environment in his study, Jersey City. Navigating the web we have often felt like one of the Jersey City residents Lynch quotes: "It's much the same all over…it's more or less just commonness to me. I mean, when I go up and down the streets, it's more or less the same thing—Newark Avenue, Jackson Avenue, Bergen Avenue. I mean sometimes you can't decide which avenue you want to go on, because they're more or less just the same; there's nothing to differentiate them" (*IC* 31). ColorColony is an attempt to address this disorganized sameness through collections of websites organized around different identities or characters represented by color. Each collection or colony is spatially rendered as a grid within which there are clusters of similar sites that collect around hubs.

While there are different organizing factors at play in each of these interfaces— information networks, imagined real estate, physical geography, or abstract colors—they all employ spatial elements that work to create a continuous and communicable common ground. It is this common ground that I see as the crucial potential introduced by interfaces that allow us to envision a strong environmental image. This common ground allows us to define our own alternatives when searching for a destination but also allows us the stability to imagine new and unfamiliar destinations.

In addition to the navigational agency it provides, this common ground has important cultural and historical implications as well. It allows us to form shared memories and histories attached to groupings and spaces that are larger than individual sites. It also has a political and social dimension. Without it, our virtual world is self-defined, producing isolated, narrow experiences in which we only see what we want to see. Experiences which are unpleasant or foreign are rendered invisible or absent by the very fact that we do not look for them.

To suggest that the environmental image should be the primary or only means for structuring an interface would fall prey to this same type of narrow vision. However, I think that if we continue to work in the direction that these interfaces suggest to produce interfaces that can supplement our existing ones, I can envision a resulting environment that has the same potential as Lynch's ideal city, a space that "not only offers security but also heightens the potential depth and intensity of human experience" (*IC* 5).

Endnotes

[1] Kevin Lynch, *The Image of the City* (Cambridge: MIT Press, 1960), 3. Subsequently cited in the text as *IC*.

[2] "Convera – News and Events: Press Releases: Consumer Search Engines Leave Professionals at a Loss, says Convera® Survey." Convera.com. 2006. 8 April 2007 <http://www.convera.com/news/pressrelease/?2006.12.19>.

[3] Alexander Chernev, "Product Assortment and Individual Decision Processes," *Journal of Personality and Social Psychology* 85.1 (2003), 159. Subsequently cited in the text as "PA."

[4] William Gibson, *Neuromancer* (New York: Ace Books, 1984), 51.

[5] "Second Life: Your World. Your Imagination." Secondlife.com. 2007. 9 April 2007. < http://secondlife.com/>.

Electronic Book Review 4.0: Toward a Semantic Literary Web-Based Interface

The Electronic Book Review showcases experiments in design, intellectual property, authorship, semantics, taxonomy, and reading practices.

Anne Burdick
"The Interface as a Tool for Reading and Writing: The Design of *Electronic Book Review*'s Graphical User Interface"

Joseph Tabbi
"All Over Writing: The Electronic Book Review (version 4.0)"

Ewan Branda
"A Map of Relations: Three Interfaces in the Electronic Book Review"

The Interface as a Tool for Reading and Writing: The Design of *Electronic Book Review*'s Graphical User Interface

Anne Burdick

Introduction:
A Distinct Kind of Reading and Writing

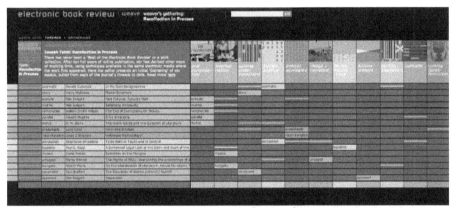

http://www.electronicbookreview.com/action/Weave?gatheringId=12

Our panel is comprised of three different disciplinary perspectives on the application design of Electronic Book Review (***ebr***): that of myself as the interface designer; Ewan Branda, the database designer and programmer; and Joseph Tabbi, the editor. It is the tension of our tripartite collaboration – more than the cooperation – that has given shape to the deep structures of the journal.

My paper begins where the reader's experience begins – with ***ebr***'s graphical user interface – which I will discuss as a spatialized writing (and reading) environment that shows how design and writing are inextricably bound in the site's visual weave. Ewan Branda will show how the logic of the interface relates to the systems of meaning made possible by the site's technical infrastructure, an extension of his own work in architecture and informatics. Joe Tabbi will discuss the media-specificity of what he calls the "all-over writing" of the site as part of the larger project of building the field of literary arts in a new media context. Our individual perspectives meet where the interface makes possible a very distinct kind of reading and writing.

Writing Space Design

In my own work as a designer, I collaborate with editors, writers and texts performing what I'd call *writing space design*, a reference to Jay Bolter's notion of a writing space in his book of the same name. According to Bolter, "the organization of writing, the style of writing, the expectations of the reader—all these are affected by the physical space the text occupies." (85) Shifting away from the idea of designing "books" or "websites," instead I design *spaces for writing* whose material composition is integral to the writing strategies and semantic outcomes of a text.

In addition to Bolter, I draw from a wide range of references within both literature and visual culture. I am exploring how communication can change when scholarly writing engages with visualization strategies borrowed from maps and diagrams, comic books, e-mail, computational design, graphic novels, and the like. In addition, I'm interested in what is possible when both analog and digital spaces for writing are designed not in the service of writing but in a dialectic interplay with it, much like program and floor plan in architecture.

In this paper, I will provide an overview of ***ebr*** 4.1 which is an ongoing experiment with these ideas. But first I want to explore two very different collaborative projects that demonstrate this interrelationship between writing and design.

Design as an editorial activity:
The structure of the *Fackel Wörterbuch: Redensarten*

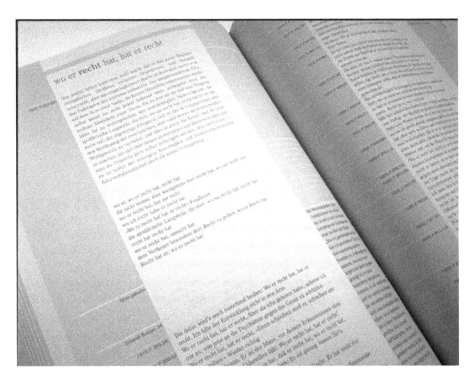

The *Fackel Wörterbuch: Redensarten* is the first of three text-dictionaries whose corpus is Karl Kraus's journal *Die Fackel* which was published in the early 1900s in Vienna. Fackellex 1, as we call it, is a dictionary of idioms.

In the dictionary there are only 144 entries and over 1,000 pages in the book. Working with the specificity of the original corpus material – which used typography and layout in unique ways – I developed a three-columned diagrammatic display that includes a "spine" running down the center of each page in order to accommodate photographic reproductions of entire pages of *Die Fackel*. On the left-hand side of the page are the documentation texts – those that quantify and categorize. On the right-hand side are the interpretive texts – the editorial glosses and commentary.

The size and position of the outside columns determines the form of the writing – its length, format, and internal composition – at the same time that it makes possible the interplay between texts through contiguity, juxtaposition, distance, sequence, and other forms of visual rhetoric.

Composite reading and writing:
Visual quotations in *Writing Machines*

In the book *Writing Machines*, Kate Hayles analyzes three works: *A Humument* by Tom Phillips, *Lexia to Perplexia* by Talan Memmott, and *House of Leaves* by Mark Danielewski. Kate's working manuscript was filled with descriptive passages and quotations drawn from these sources. Since Kate was interrogating the materiality of the original works, I suggested that we use a form of visual-verbal quotation – pictures of the original texts intertwined directly into her text in order to capture as much of the materiality of the originals as possible. The resultant design/writing strategy was described by the futureofthebook.com as "a new, composite reading mode" that is both viewed and read by a skilled reader.

The writing space design changed the character of the reading and the writing. The visual-verbal quotations communicate on multiple levels, thereby reducing the need for cumbersome verbal description, an outcome that mirrors the influence of photography on art and literature a century earlier.

Media-specific designing and writing:
The writing spaces of *Electronic Book Review*

Electronic Book Review 2.0
In the early days of *ebr*, we were interested in creating forms of writing that had no print corollary, writing that was structured to perform in ways that only digital writing can. These experiments were much like those that we now find in *Vectors* (www.vectorsjournal.org) – standalone essays that were mini-sites unto themselves.

Here is a detail from the home page from *ebr*, version 2.0, when the site was comprised of single-themed issues published at regular intervals.

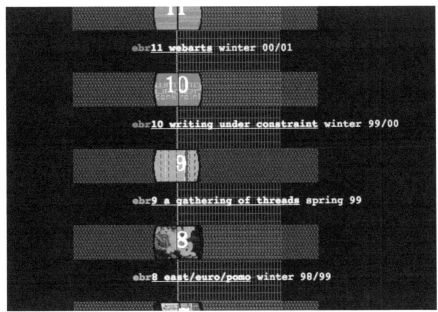

http://www.altx.com/ebr/threads/threads.htm

The contents page for **ebr** 9 shows evidence of the breakdown of the discrete thematic units that were built into the site's interface and editorial strategy. **ebr** 9 is a gathering of themes (**THREADS**) found in previous issues, an indication that the publishing model had run its course.

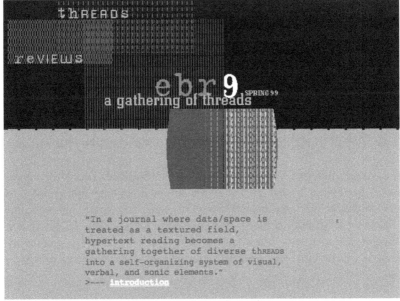

http://www.altx.com/ebr/ebr9/index.html#

http://www.altx.com/ebr/ebr9/index.html#

Pre/post-erous: La Jetée Ciné-Roman is an example of one of our design-writing experiments from 1999 by Tracy Biga Maclean, Chris Peters, Jon Wagner, and designer Sophie Dobrigkeit. It opens into its own distinct window and juxtaposes text and images to compose its critique of the book under consideration. The non-linear structure leads to a collaged reading experience that would be difficult to translate into print.

Electronic Book Review 4.0

In the early 2000s, it became clear that while we were engaging the medium on an essay-by-essay basis, the journal itself was still tied to a print paradigm of regularly scheduled publication and single-themed issues.

We moved toward a conception of the journal as a living archive in which old and new writing projects could be drawn together and remixed according to the interests of *ebr*'s community and to relatedness determined by the system.

At the same time we were interested in autopoeisis and the unexpected outcomes generated by an interface that was a literal meeting point between reader's interests and editorial activities. In lieu of a single, all-encompassing map or view of the site's contents, we developed an interface display that would reconfigure the contents according to the discursive activity unique to that moment in time. And, importantly, the displays were designed as spaces for writing.

HOME PAGE

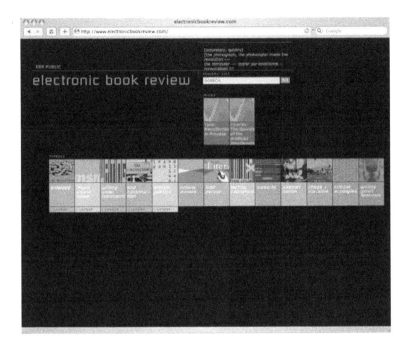

The home page is organized according to chronology. The editorial activity runs along a horizontal axis from most recent on the left to the oldest on the right.

The scrolling text at the top of the page is the most recent e-mail announcement from **ebr**'s **BARKER**. This is the equivalent of "latest news" or "now hear this…" Below the **BARK** are three points of entry to **ebr** content which correspond to the three groups that comprise the **ebr** community.

(1) The **READER'S LIST** is a search function that generates a mix of contents according to a reader's interests.

(2) The **MIXES** are curated sets of essays that are created by a community that we call **WEAVERS**. Weavers are invited participants who can gloss and remix the site's contents.

(3) The **THREADS** are ongoing themes initiated by the editor. Each thread has its own color and icon. Each essay that is added to the database is entered within what we call a primary thread, which is its original home. But to encourage lateral movement and connectivity between documents, essays may also have secondary affiliations with other threads.

Beneath the thread icons are pull-down lists of the most recently active articles within each thread, going back six months. "Active" means either newly published or recently commented upon, responded to, or added to a mix by a weaver.

....

From the home page you can easily go to the **THREAD PAGE** by clicking on its icon.

THREAD PAGE

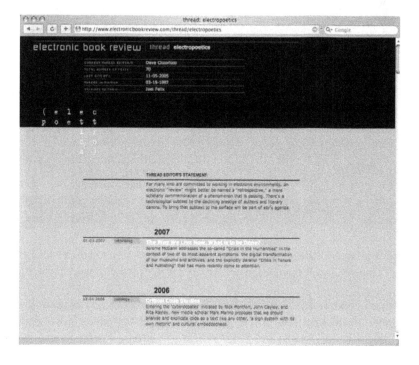

On the thread page you can see the entire history of a single thread (in this case for "Electropoetics") presented in a conventional table of contents format that moves back in time as you scroll down the page.

To the left of each essay is a small rectangle with a 13-letter code. We call this device a **TEXTCODE**. This code is a kind of linguistic icon, a visual-verbal marker that stands in for an essay in a variety of context-specific visual mappings that appear throughout the site.

....

Clicking on the journal name in the upper left-hand corner sends you back to the home page.

HOME PAGE → SEARCH: ESKELINEN

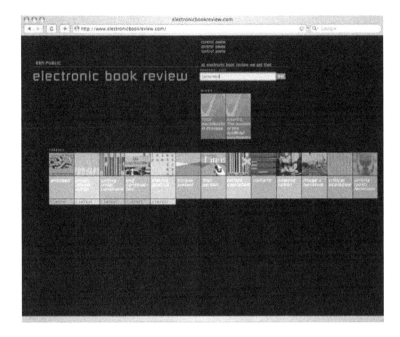

From the home page we can enter the search term "Eskelinen" in the search box to generate a reader's list.

WEAVE: READER'S LIST → SEARCH RESULTS: ESKELINEN

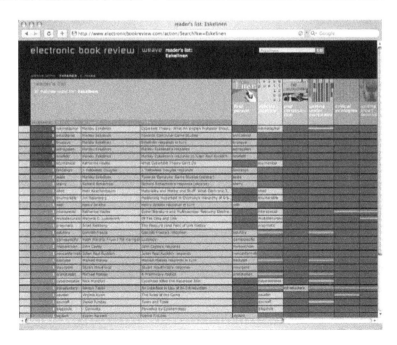

The reader's list is displayed within the **WEAVE PAGE.** The search results appear within a colored grid that lists the relevance ranking, textcode, author name, and title for each essay that is called up within the results. To the right of the list, a visual "weave" is generated horizontally by those threads with which the essays are affiliated. As you move to the right, you can see how each essay has a **PRIMARY AFFILIATION** – indicated by the textcode in a colored box – and a **SECONDARY AFFILIATION** indicated by a narrower strip of color. The visual mapping brings together decisions made by the user, writing by the author, and the editorial activity of the system.

From here the reader has a range of navigational choices: she can select an individual essay to read or move to a thread page using the thread icons.

This display gives the reader a sense of where the writer Markku Eskelinen fits in the world of *ebr* through the visual information – the color, the language of the textcodes, the sequence of threads, and the frequency of textcodes and other marks within each thread column.
....
Typing "Amerika" into the search bar will demonstrate how the appearance of the weave changes in response to the list of essays.

WEAVE: READER'S LIST → SEARCH RESULTS: AMERIKA

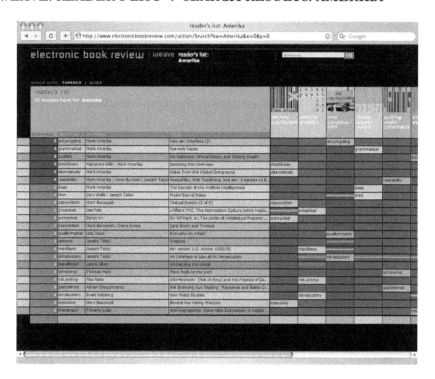

It becomes instantly apparent that Mark Amerika's contributions are more wide-ranging thematically than were Eskelinen's.
………………………………………………..
Searching for Friedrich Kittler shows a different result.

WEAVE: READER'S LIST → SEARCH RESULTS: KITTLER

Within each weave display, the reader can choose to **WEAVE** the essay list with either threads or mixes. The default is to weave with threads, which has been demonstrated so far.
………………………………………………..
Clicking on "mixes" in the upper left-hand corner shows how this list of essays intersects with the interests of *ebr*'s weavers.

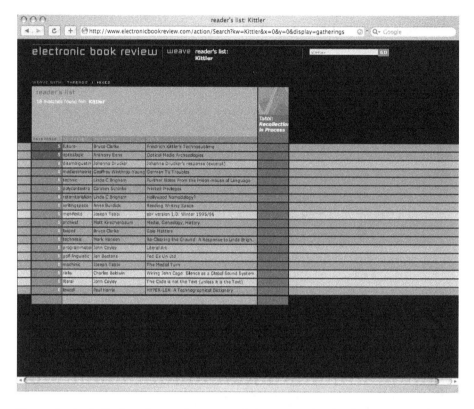

The refreshed display shows that one of the essays in the list is a part of Joe Tabbi's mix called "Recollection in Process."

....

This allows us to click on the icon for "Recollection in Process" to view the contents of Tabbi's mix which will be displayed within the weave page.

WEAVE: MIX → RECOLLECTION IN PROCESS

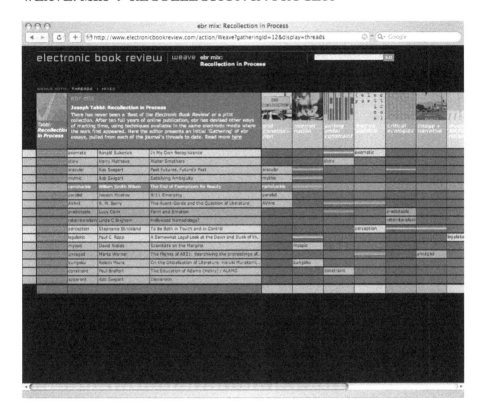

Here you can see that a mix appears within the weave display in a manner similar to that of a reader's list. In the left-hand column you see the list of essays that Tabbi curated from the database of *ebr*. Each essay brings the textcode, author name, and affiliated threads into the weave accordingly.

Members of *ebr*'s community of weavers can use the site's back-end tools to create their own mix. (Since this is a new feature there are relatively few at the moment.) The weaver names their mix, writes an introduction and a blurb, selects a set of essays, and saves it all to the database. The mix automatically appears at the appropriate places throughout the site according to the logic of the system.

....

So far we have looked at the systems of organization at the level of the journal. Now let's look at an essay. Clicking on an essay title or textcode within the weave will take us to what we call a **TEXT PAGE**.

TEXT PAGE: [ramshackle] WILLIAM SMITH WILSON

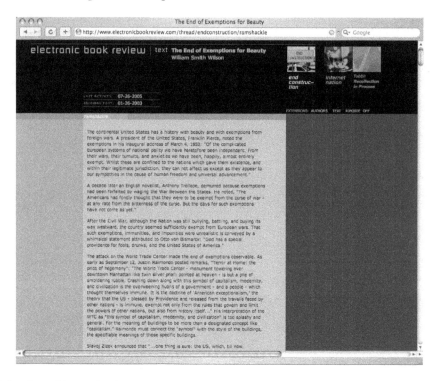

The thread icons in the upper right hand corner of this essay by William Smith Wilson show the essay's primary and secondary affiliations. The primary affiliation shown here is "End Construction." This thread affiliation determines the color fields within which the essay is displayed. You can also see that this text has a secondary affiliation, "Internet Nation" and that it is also a part of Tabbi's mix: Recollection in Process. The reader can click on these icons to open the corresponding thread pages or weave.

On a text page the reader can move within and between texts in two ways: from inside the main text through hyperlinks and through the marginalia that can be displayed in the columns on either side.

In the left-hand margin, weavers can write commentary which readers can view by clicking on links from inside the main text. Ewan Branda will cover what we call the **GLOSSING** function in his paper.

The text's **EXTENSIONS** – displayed in the right-hand margin – enable movement laterally across the database from essay to essay. A variety of options is automatically generated by the system that is meant to allow movement in the act of reading.

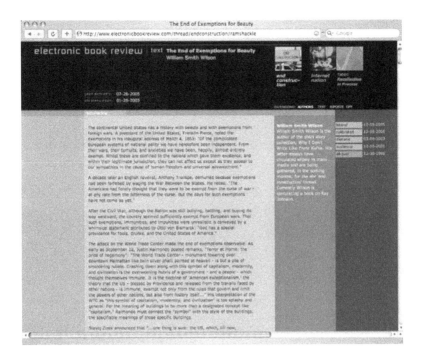

Clicking on the **AUTHOR/S EXTENSIONS** link reveals an author's bio and includes textcodes for other essays by the same author that can be found within *ebr*. Clicking on a textcode opens that essay.

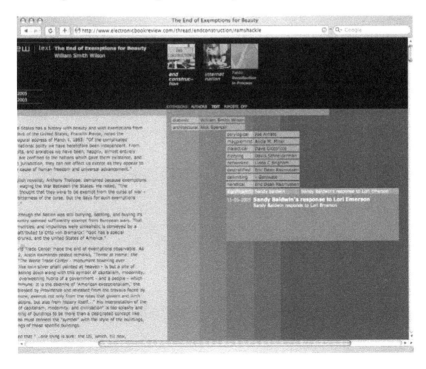

The **TEXT EXTENSIONS** link displays other texts within the database that have an affinity with the main text. Like the weave page, this display uses color, proximity, and position to represent the metaphorical closeness or distance between essays. Essays themselves are represented metonymically with author names and textcodes, creating a kind of topographical shorthand – or spatialized written representation of the relationships.

When the reader wishes to make a choice, the textcodes can be rolled over to reveal the full title and essay blurb. Clicking on a textcode opens that essay.

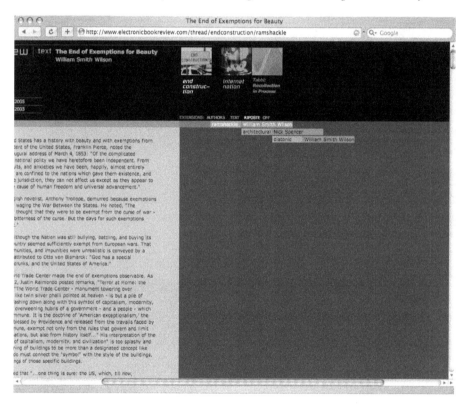

The **RIPOSTE EXTENSIONS** link reveals textcodes and author names for essay-length responses to Wilson's essay. Within *ebr*, writers can respond to texts in two ways: through paragraph-length **GLOSSES** in the margins or with longer texts called **RIPOSTES**. The riposte link shows up only when an essay has a riposte or is part of a **RIPOSTE CHAIN**. So you can see here that there are two ripostes, an exchange between Wilson and Nick Spencer.

....

Clicking on the textcode for Spencer opens his riposte to Wilson.

TEXT PAGE: [architectural] NICK SPENCER

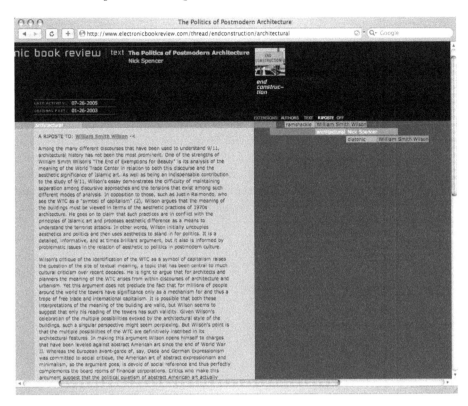

Within the riposte extension in Spencer's essay you can see the position his riposte occupies within the riposte chain. The display allows the reader to follow the discussion in either direction by clicking on the textcodes.

Conclusion: Designing and Writing

As this demonstration shows, the interface marries the signifying system of the diagram (position, color, scale) with written language in the form of names, titles, blurbs, and textcodes. The graphical user interface uses imagery and visual iconography in very small doses because we wanted to maintain an emphasis on reading, and by extension, on writing.

The textcodes in the weave and the text extensions are simultaneously symbolic and indexical in the Peircian sense, for they can be read as texts, seen as representations or codes, and can be used as navigational handles. The glosses, as Ewan will demonstrate, are a kind of situation-specific dialogic writing: a practice enabled by the features of the interface and the system. Hence *ebr*'s interface can be understood as a diagrammatic, distributed, all-over writing space.

As a custom writing and publishing application, **ebr** is distinct from word processing, wiki, and blogging software such as Typepad in which "design" is basically a change of skin, for at **ebr** the design is integral. But the most significant difference is also the least visibly apparent: the writer-user of a commercial software package is confined by compositional structures whose conventions are somewhat transparent and are by definition generic.

Not so with **ebr**: when editorial activities rub up against the database structures which in turn strain to connect with the visual mapping of the interface – or vice versa – the instrumentality of the site design's boundaries, rules, and spaces are revealed. The chafing tells us that it is time to carve out a new kind of space or shift the rules of engagement. Within this ongoing, unfolding, collaborative project, it is the push and pull between the writing done *inside* the system and the designing of the structures of the system that generates **ebr**'s unique visual and linguistic form.

All Over Writing: <u>The Electronic Book Review</u> (version 4.0)

Joseph Tabbi

Introducing a collection of scholarly essays, <u>Debating World Literature</u> (Durham: Duke University Press, 2004), Christopher Prendergast cites an observation by Arjun Appadurai that should give pause to anyone who wants to create a space for literature in new media: "public spheres," Appadurai writes, are "increasingly dominated by electronic media (and thus delinked from the capacity to read and write)." [22] That "thus" can rankle. Obviously Appadurai is not thinking of the Internet, which is still (and likely always to be[1]) overwhelmingly textual, despite an increasing visual and insistently instrumental presence. The assumption that reading and writing are of course "delinked" from all electronic media, shows just how deep the separation of spheres has become for scholars in the field of post-colonial cultural studies. Any notion that electronic literature might in fact <u>be</u> an emerging world literature is foreclosed at the start.

It wasn't supposed to be like this. Appadurai's casual dismissal of reading and writing as active elements in "electronic media" should seem strange, if one recalls the idea advanced by cyberculture visionaries for a universally accessible, open-ended archive primarily for <u>texts</u>. That was the idea behind Vannevar Bush's <u>Memex</u> and Ted Nelson's "hypertext" – not the current expanse of decontextualized "hot links" but rather a way of bringing documents, in part or in their entirety, to a single writing space for further commentary and the development of conceptual connections. Another word Nelson coined for the process was "transclusion" – an inclusion through site transfers that could be full or partial, depending on one's requirements: in every case, the "original" document remains at its home address while being reproduced at the target address (not just referenced or linked). The achievement of this capacity, which can make reading and researching also a kind of worldwide consortium building, brings to the public activities that had been considered, like much of print culture, private and secluded.[2] Realizing such a collaborative network in the field of literary scholarship is behind the current version of <u>Electronic Book Review</u>. The essays by Anne Burdick and Ewan Branda describe how the interface works, technically and from a design point of view. Here, I discuss how the interface might be <u>made</u> to work in the transformation of critical writing. Electronic interfacing, as practiced since the implementation of <u>ebr 4.0</u> (early in the year 2007), has a chance to bring a distinctively <u>literary</u> practice back into the operational field of computing and text processing. Connections that over time have become, in print, conceptual and implicit, become explicit and readable not through technical means alone (e.g., the "hot link"), but by the strategic placement of words, sentences, and other semantic elements in every space afforded by the screen. Even the URL of an ebr publication says something, not only about the electronic address of an essay, a narrative, or an essay-narrative, but about its content; literary concepts are "tagged" in each essay, and the tags are developed in awareness of keyword and metatag development at affiliated sites throughout the Web. Though possible and, in our view, desirable, transformations in the practice of critical writing are by no means inevitable and they will depend, not on <u>ebr</u> or any one site, but on the development of a consortium of sites and a consensus about "best practices" that answer to, and can help direct, practices under development in <u>ebr</u>.

The creation of the <u>ebr</u> writing space, even as it looks back to Ted Nelson, also looks forward to another, as yet unrealized, conception of knowledge processing on the Internet – namely, the Semantic Web. The Semantic Web is most useful as a metaphor at this point, since its realization depends not on a top-down development but on the independent decisions by many site developers to mark up and tag text according to a common and communicating set of references.[3] My interest in the Semantic Web is its potential, through the mundane task of tagging documents, for developing not only a database but a vocabulary specific to the field of e-lit, using procedures that involve both ordinary readers and editor/curators. At the same time, the necessary awareness of vocabularies under development elsewhere must not influence the autonomous development of a literary metatag vocabulary for literary purposes. In whatever ways the literary field is transformed by electronic environments, its transformations should be readable in terms created by, and for, literary authors.

Certainly, its creators want to make the conceptual writing in <u>ebr</u> consistent with the predominant flow of information among sites whose developers recognize the need for pooling content. Nonetheless, editors and authors cannot assume that our attention to "semantic" content will be enough to sustain a literary presence on the Internet. Where tagging and linking depend on direct, imposed connectivity at the level of the signifier, the creation of literary value depends on suggestiveness, associative thought, ambiguity in expression and intent, fuzzy logic, and verbal resonance (where slight differences, not identifications among fixities, are the origin of meaning – "the difference that makes a difference," in Gregory Bateson's phrase; Emily Dickinson's expression of "internal difference / where the meanings are," the "topologies" that, according to Michel Serres, "haunt" the geometries where most people live).

Tags are important; naming is one of the literary arts. But the names need to change, new names need continually to be created so that the tags read by machines do not appear (to living readers) as word soup. The need to combine this literary development with the machine-readable content that would characterize an operative Semantic Web is a challenge not only for ebr but for any site interested in knowledge creation that depends on, but is never identical to, information storage and retrieval.

All Over Writing

At a time when powerful and enforced combinations of image and text threaten to obscure the differential and processual ground of meaning, ebr seeks to recognize and encourage the potential for bringing together, rather than separating, rhetorical modes in the production of nuanced, textured languages within electronic environments. Much of what we present, online, is recognizable from the tradition of print: the self-standing essay, the book review, editorials, descriptive blurbs, and so forth. What distinguishes our presentation from print, however, is a way of linking content together through conceptual writing, so that relations that tend to be implicit in a print archive are made explicit and present in one place. Following a reference or an allusion or even a hint, readers needn't go to a different bookshelf, library, or archive. The term I want to offer, for such a critical enterprise, is drawn from the arts: bearing in mind the "all over painting" in abstract expressionism, I want to propose an "all over writing" that embraces seriality and interconnectivity, rather than being distracted by links. It happens that this term, "all over painting," figured in my first book, on relations of technology and contemporary fiction, which was published at about the time when I conceived The Electronic Book Review. For documentary purposes, as well as for purposes of visual illustration, I will launch this discussion with a brief reference to my book – or rather, to the cover (see Figure 1).

Reproduced here is "Small Higher Valley 1" (1991), the first in a series of paintings by the New York based poet and painter Marjorie Welish. In the book's introduction, I described this painting's use of "a virtual system of geometric sectionings to suggest the networks and grids that underlie rational thought," while at the same time avoiding any single total system that could dominate everything (Postmodern Sublime, 20) That description resonated with my topic, the sublime in American fiction as it was finding expression in a group of authors whose work registers the emergence, post-World War II, of technologies of information, communication, and control. Only recently, while unpacking my library in the Summer of 2007, did I happen to notice how similar in some ways Welish's multi-colored grids are to the visual design of the ebr weave by Anne Burdick: http://www.electronicbookreview.com/

Figure 1: _Postmodern Sublime_ (*Ithaca, NY: Cornell University Press, 1995*)

Notice for example this page's capacity to include any color against the generous black background, its flexibility and expansiveness made possible, not limited, by material constraints of the line and the screen. The grid gives precise and measurable locations – they are the known habitations for both the viewer of Welish's painting, and the user of interfaces. But, to cite Serres again, one can live in geometry and still be haunted by topology. The grids also serve to stage a sequence of wholly <u>relational</u> meanings. Paint is allowed to brush or bleed into the adjoining quadrilateral sector (in the Welish painting); while files, placed under columns and listed chronologically, also radiate outward to other files based not only on informational content, but on conceptual similarities that might be recognized by readers and editors, though not necessarily anticipated by authors. The design is relational and open-ended, and the electronic writing space is extended "all over," so that one site or essay can be included within and transferred to other sites.

Market

In my talk today, I want to discuss some of the ways that our mode of "all-over writing" capitalizes on what the Web allows, enabling a media-specific reading and writing practice. But to understand this specificity and why it requires designers and programmers working with, but necessarily independent of, writers, I need to say a few words about what our journal is <u>not</u> trying to do. First of all, we're not competing with print, and we're not trying to reproduce the traditional peer review academic journal (see "coda" below) or the just-in-time delivery of established review media. Though it sounds odd to say it, even slack – there's no reason that an essay or book review needs to appear close in time to the texts under discussion, except for the commercial (and relatively recent) enforcement of brief shelf lives for books on the one hand, and platform obsolescence for most Internet sites on the other hand. Those limitations are not inherent in books or websites. Obsolescence – a theme in <u>ebr</u>'s earliest manifesto – is not a technical problem, but a political and economic one:

> Tabbi, Joseph. *"A Review of Books in the Age of their Technological Obsolescence." <u>ebr</u> 12-30 1995.*
> *http://www.electronicbookreview.com/thread/electropoetics/manifesto*

With the rise of neo-liberal economics in the1980s and 1990s came a consolidation of major book publishers, a proliferation of small presses even as local bookstores were in decline. Authors were being transformed into performers and book peddlers, and the desktop itself was being converted into an environment suitable primarily for office work and innovations in marketing. In general, such transformations have been catastrophic for reading and writing, as my opening quotation from <u>Debating World Literature</u> suggests. The Web's been around long enough, that one can safely say that older, bounded, forms of the literary are not likely to re-appear in the current, user-friendly environments. There is of course a wealth of experimental, non-narrative fiction and poetry in non-commercial platforms. But if we haven't had major born digital novels or poems by now, probably we never will.

In my own practice of critical writing within English and Arts programs in the United States, these developments would seem to be consistent with the rise of cultural studies and media studies – where graduate students, instead of taking the years and sometimes decades needed for mastering a subject, are encouraged to publish even as they are still taking courses, carrying a teaching load, and often holding a job. There are now, at my State University in Chicago, even academic conferences for undergraduates. Literary and Cultural journals on the Internet tend to be short-lived, and platform obsolescence has made it difficult to establish the canon of contemporary texts that is necessary to the sustained critical discussions needed to form a field. The literary work produced in such a climate has been characterized, in an essay presented to the German "Network" on "American Studies as Media Studies" by the critic John Durham Peters, as a kind of "just-in-time production." With low start-up costs and low barriers to entry in terms of knowledge, and the ability to "supply the increasingly important cultural industries with savvy employees," the new curricula can be recognized as a kind of "academic parallel to new-liberal economic policies." Such curricula, and the mostly student-run Internet journals that support them, often encourage a topical, informatic approach to scholarship that might be summarized in the formula: "find a hot topic, add theory, present paper."

> *Peters, John Durham. "Strange Sympathies: Horizons of Media Theory in America and Germany." Paper presented at conference of the Deutsche Gesellshaft für Amerikastudien, "American Studies as Media Studies," Gottingen, 10 June 2006. ebr 03-01-2008*
> *http://www.electronicbookreview.com/thread/criticalecologies/justintime*

If the Internet were just a way of making that process still more efficient, I would have left the field years ago. In fact, all but a few of my colleagues in Literary Criticism, Theory, Fiction and Poetry Writing, have left, or relocated to departments of New Media, Arts, and Communications. After establishing a career where reading, writing, and traveling are the primary activities, what author would compromise such autonomy for a career of Project Management, Grant writing, long-distance and frequent commuting to corporate conferences, and continual subjection to programs and platforms that routinely confuse commercial interruption and technical instruction? As Linda Brigham writes in her ebr review of N Katherine Hayles, there is something abject about our dependence on expensive, controlled goods, and even the celebrated distribution of agency in networks has its limits: "Feedback to network nodes," Brigham writes, "seldom indicates the nature of the network; that information yields only to a higher level of surveillance and analysis, while the nature of the network feeds some entity beyond us, we continue to subsist on the empty calories of ideas and concepts." A similar note is sounded by Andrew McMurray, in his introduction to our Critical Ecologies thread: the idea that everything, even literature, needs to be done using computers, might serve the current technocracy but it has mostly rendered transactions and communications "sclerotic."

Brigham, Linda. "Do Androids Dream of Electronic Mothers" ebr 11-09-2006
http://www.electronicbookreview.com/thread/electropoetics/liminal

McMurray, Andrew. "Critical Ecologies: Ten Years Later" ebr 12-01-2006
http://www.electronicbookreview.com/thread/electropoetics/ecocritical

Unforgiving as this critique may be – and I agree with it – the work of McMurray, Brigham, Peters, and many others writing for ebr also indicates a way forward: these writers, after all, are not only offering critiques; and neither are they simply transcribing their critical writing from one medium to another. What they are doing, in most cases, is reflecting on the medium and their own relation to networks as they join with (or engage in principled argument against) other writers within a network that is identified in the process of writing. This engagement involves more than an adjustment of attitude or achievement of competence with computers and databases. To subsist on more than concepts, one needs to bring one's own work into contact with other, related work, so as to be recognized by others who, writing critically in ebr and elsewhere, have made similar recognitions on the basis of involvement in similar projects, similar discussions. Every technical innovation in ebr, fundamentally, is geared toward the realization of this one goal: to bring the electronic network and its nature into consciousness. What we are working toward is the possibility not simply of literature's inhabiting networks, but for literature to become a network.

Emergence

Once that goal is recognized, it becomes possible to imagine a place for doing the work of literature without expecting miracles, revolutions, or the end of books. I refer to "the work of literature," not "works" of literature, for a reason: namely, talking about processes makes more sense in electronic environments than talking about objects, even when the objects are verbally inventive and could only be devised using new media. At ebr, in the threads titled Webarts and Image + Narrative, we give extensive coverage to conceptual and literary arts that explore their newfound media specificity, but we're not a free-standing art project. ebr accommodates, but does not encourage, critical hypertexts and other self-contained, custom projects because these tend to proliferate connections internally, encouraging reading in isolation.

What we are trying to do at ebr is to develop and maintain an advanced literary culture within the new media. There's an aesthetic, over the years, that ebr has advanced fairly consistently, and it can be seen in the examples that Anne Burdick and Ewan Branda have on display in their essays (appearing jointly with this one). I, together with my co-editors at the University of Illinois at Chicago, New Zealand, SUNY Buffalo, Leuven, Boulder, Colorado, Siegen, Germany, Munich, Atlanta, Georgia, and elsewhere, have tried to express the nature of that aesthetic in editorials and in comments seeded throughout <u>ebr</u> in blurbs, glosses, essay id's, and other small, para-textual elements that can be viewed in the screen shots in the accompanying essays by Burdick and Branda. The "all-over" aesthetic can also be sensed visually, in the non-verbal judgments implied in gatherings, threads, and folds; and it can be generated as much by the database structure as by content.

Knowledge, in such an all-over writing environment, is produced not directly, but as a meta-phenomenon, traceable to (though not identical with) the tags, keywords, and descriptors that authors use (or that they leave to the programs they're using, in which case authors cede more autonomy to the machine than they might know). Here, the actual knowledge is not produced, not entirely, by the content of an essay but is "put in" by the author or editor, and the uses of such knowledge are not realized until a reader enters the picture, following a gloss, or connecting one tag with another, identical or related tag. In this sense, knowledge production in a networked environment is "virtual" – which is to say, it is given as a potential, in the act of tagging, and realized only when the relations among tags are recognized by a reader or made noticeable by an editor.

In following such connections and enfoldings, the reader does not in any sense <u>replace</u> the author as a producer of knowledge; rather, the reader produces a different knowledge, constructed not only from works the reader has read, but from the works' self-descriptions. The knowledge is, from the very start, already <u>relational</u>: and this is what makes it appropriate for a networked environment driven as much by semantic encoding (regarding what works are <u>about</u>) as they are by syntactic and structural coding (regarding what the works are made of materially – its letters, sentences, and so forth).

Jerome McGann, in his <u>ebr</u> essay on the electronic future of the Humanities, mentions in passing the "severe critique of critique from what D. G. Rosetti called 'an inner standing-point' – that most telling of critical positions." McGann, of course, creates his own set of critical references, including his near-contemporary, Bruno Latour, as well as past self-critical critics, creators, and philosophers such as Rosetti and Alfred North Whitehead. That is what any scholar must do, in addressing him or herself to peers in a literary essay. But, in addition to the author's self-chosen references, the gloss on Rosseti takes readers to a critique of McGann's own Rosetti Archive, by Katherine Acheson in <u>ebr</u>. Still further, but invisibly, the term, from "an inner standing-point" has been tagged with the keyword, "focalization." And so the entire essay is not just linked notionally to essays on or by Rossetti, McGann, and the field McGann consciously enters; McGann's work also has been gathered, through the database, to literary works (for example, Rob Swigart's short story, "Dispersion") that experiment with focalization as well as several essays that discuss the concept critically. Further still, once the tag is in place, it will be linked automatically to future works on that topic, as they are recognized and tagged by future ebr editors.

Acheson, Katherine. "Multimedia Textuality; or, an Oxymoron for the Present." ebr 11-11-2006.
http://www.electronicbookreview.com/thread/criticalecologies/illuminated

Swigart, Rob. "Dispersion." ebr 10-27-2006.
http://www.electronicbookreview.com/thread/fictionspresent/apparent

Truly, if there is such a thing as "all-over writing," it cannot be defined by pointing to specific features or information; the site needs to be worked with, read, so that significances that are relational have time to emerge. (A similar, serial effect is given in all-over painting: the "Small" and the "Higher" in Welish's "Valleys" would be impossible to discern on just one canvas from this series: smaller, higher, than <u>what</u>? <u>Where</u>? Such questions are meaningful only with regard to relations that are produced as the painting is created, and as the painter's decisions are recognized by viewers over an extended time of viewing.) Emergence does not produce an object; meaning in online writing cannot be traced or reconstructed by monitoring hits or reader trajectories, meaning can only be held in mind, while reading, writing, or gathering essays onsite. I'll give here two examples – both of them conventional enough to look at, but connected in ways that are recognizable in the process of reading the essay. In doing this, I present <u>ebr</u>'s first "enfolded" site – a project description at the University of Virginia "NINES" website that is coherent with what's been happening at <u>ebr</u>. Rather than simply "linking" to this site, we've brought the essays in their entirety from the NINES site into <u>ebr</u>. The essay remains on the NINES site, but its description, its metadata pointers, are brought into the <u>ebr</u> database. The essay itself is, in a sense, "wrapped," so that (from a reader's point of view) it is as much a part of <u>ebr</u> as the essay by Jerome McGann which mentions the project (a "Networked Infrastructure for Nineteenth-century Electronic Scholarship").

McGann, Jerome. "The Way We Live Now, What is to be Done." ebr 01-03-2007
http://www.electronicbookreview.com/thread/electropoetics/rethinking

The NINES and COLLEX projects, referenced by McGann in <u>ebr</u>, can be accessed on the University of Virginia Web Site as well as at <u>ebr</u>:

Jerome McGann and Bethany Nowisky. "NINES: A federated model for integrating digital scholarship." ebr 04-09-2007
http://www.electronicbookreview.com/thread/enfolded/collaborative

The blurb leading to this essay stresses the coherence (and difference) between MgGann's and Nowisky's project and the overall ("all-over") ambition of the

<u>ebr</u> interface, namely:

> *NINES is an initiative at the University of Virginia to "establish a coordinated network of peer-reviewed content and tools." We present the project here because it's consistent with the initiative at ebr to create a peer-to-peer literary network for conceptual writing.*

This example is meant to demonstrate the reach of the <u>ebr</u> interface, a model for collaborative reading, and a mode of collaboration among sites that has been, too often, forgotten by busy-bee writers. The ease of "linking" makes it unlikely that editors will consider, in detail, what arguments, keywords, metatags, and implied audiences essays from the two sites might have in common. That such collaboration demands explicit negotiation between site editors, who are expected to grant permission without seeking payment in return, is a necessary and desirable feature for the construction of a literary network. It carries into the new media one aspect of scholarly interaction that critics and writers cannot readily do without, namely, the gift economy among literary and cultural peers.

As I would not want to imply that <u>ebr</u> could achieve its "all-over" ambition by itself, I should also mention the consistency of our project with other advanced sites (such as the <u>Archiving the Avant Garde</u> Project , *NT2 : Nouvelles textualités, nouvelles technologies*, and the proposed ELO Directory.). Such sites are not dedicated to the advancement of one specialized discourse; they, too, are about building a field, and creating a context for the persistence of literary and conceptual arts in new media. All such projects need to be developed in awareness of each other and the vocabularies being developed in many nodes, but capable of being gathered universally into a Semantic Literary Web. Whether or not the SW becomes a reality, it offers a good point of reference and a general direction for our project as it might connect with other projects.

Coda: Peer to Peer

No one, I think, will dispute the desirability of developing Semantic Web standards that are suitable for literature. Few ought to object in principle to establishing consortia of mutually recognized sites so that a vocabulary standard can evolve over time and under a range of institutional contexts. For such cooperation to gain traction, however, mechanisms of review need to evolve along with the standards. These review mechanisms, to be more than privileged community gates, also need to be in place universally, throughout the literary profession. In the past, at universities worldwide, the "peer review" system has developed in response to this need for standards, which is in reality twofold: 1) to keep track of terminology and conceptual trends so as not to turn the ivory tower into a tower of babble; but also, 2) to uphold standards of quality. The communicative function, which is managerial, is not always conducive to the qualitative function, which is a matter of agreement and disagreement among many subjectivities, among authors, readers, and (most important to the field development) readers in the process of becoming authors. Bringing these two functions together, the administrative and the evaluative, is the challenge of academic review.

Beyond even these dual necessities of quality control and career advancement, peer review is also, perhaps primarily, a mechanism of sorting. It's how professionals, would-be colleagues and collaborators, select some materials for attention and reference (in the process necessarily excluding the majority of materials and producers). Only through selectivity can the efforts of professional readers and their students be responsibly marshaled. Reading lives are limited, and this material condition is what necessitates the development of literary canons and what justifies the employment of accredited professionals to teach canonical works and their differing receptions in different historical periods.

While necessary in principle, "peer review" can easily be corrupted when faculty themselves no longer have time collectively to read the work that their own profession is producing at record volumes even as tenured lines at universities worldwide have been reduced drastically. As is widely known and trenchantly reported by Marc Bousquet, in the United States today, around 75% of courses are taught by lecturers, graduate employees, and other casual or temporary workers; 25% by tenured or tenurable professors. Forty years ago the proportions were reversed.

> Bousquet, Marc. *How the University Works: Higher Education and the Low-Wage Nation*. New York: New York University Press, 2008.

It would actually not be so bad, if Bousquet were simply arguing that the academic system is currently dysfunctional. But Bousquet's point, rather, is that university administrations, with the implicit support of faculty, have shaped the system to do exactly what it is meant to do - namely, to restrict the supply of peer reviewed researchers and employ an expanding force of low-wage workers whose development is subject to sub-professional performance standards. Consistent with critiques of new liberal economics generally and the critiques in ebr by McMurray, Brigham, and Peters (cited above), Bousquet connects the "informatics" of education with the "informality" of work conditions for the majority of graduate and non-tenurable teachers. From this perspective, fears of the traditional university's being displaced by electronic regimes of "distance learning" are misplaced. The packaging of education as information has already "distanced" the majority of literary professionals from the day to day activity of their own students and colleagues even within their own departments. The restrictions on what we actually get to "review" are such that our collective work, as writers and scholars, is unknown even to ourselves.

See the Techno-Capitalism thread in ebr, co-edited by Bousquet and Katherine Wills:
http://www.electronicbookreview.com/thread/technocapitalism

The consequences of this material transformation are felt in the tenured ranks themselves in many ways. Restrictions on what can be reviewed, and on the number of professors who can do the reviewing, can in effect disqualify professors from evaluating peers responsibly and selecting the texts that will be common to our disciplines. The disqualification has more to do with limitations on time than with any active attack on "academic freedom." Speaking from my own experience, over the past several years at advisory meetings for the promotion of colleagues, I increasingly have a sense that only those assigned to report on a candidate's scholarship have read the work with any care. The majority have time only to read reports sent in by "outside readers." The premise, that there is an audience of specialists "out there," better able to analyse a work of literary writing than the colleagues in one's own department, has done more to fragment the profession than any purported tendency toward jargon or politicized language in academic writing. The conceit that there is a set of standards apart from those developed internally among a cohort of professionals, only reinforces the widespread acquiescence to the imposition of standardized testing at all levels of education.

So as to avoid this outsourcing of services that need to be performed by all, not a select few, within the literary profession, and to advance efforts at reforming the academic review process, as of February 2008 ebr has placed on its site a formal statement of our longstanding, hitherto informal, practice:

> *ebr is a journal of critical writing produced and published by writers for writers: a peer to peer modification of academic review. Each essay is reviewed by a thread editor (a tenured professor) and at least one other ebr editor. On acceptance, the essay is posted to our staging site, where it is made available for comment by our 500-plus past contributors, all of whom are published authors in print and online. Unlike academic peer reports, which are generally seen only by committees, ebr reviewer comments can be read in the margins of the essays, as "glosses." More substantial response is given in commissioned Ripostes.*

This policy is in solidarity with initiatives and institutional experiments under way as of this writing, notably at the Institute for the Future of the Book:

> *Jeffrey R. Young. "Blog Comments and Peer Review Go Head to Head to See Which Makes a Book Better." The Chronicle of Higher Education, Tuesday, January 22, 2008.*
> *http://chronicle.com/free/2008/01/1322n.htm (for subscribers only)*

More generally, the development of a web-based reading culture promises to bring to academia and its publishing institutions something that has been languishing in print culture for a long time, namely: a practice where works are not only read but our various readings are recorded, and that record is itself made public. Like the standardized tests that can account only for what can be tested, standard accounts of reading can account only for elements that can be measured: in surveys giving the number of books in circulation, the time that students or teachers claim to spend reading, and so forth. If instead of measuring what is measurable, we make visible the active and participatory reading that is actually going on in our profession, we improve our chances of justifying the actual work that creative writers and literary scholars are engaged in. What we bring to our respective desktops, and what we do with the materials that arrive there, is the essence of literary work. The activitation of this process, and the case by case transclusion of work by our self-selected colleagues, is not just a realization of the technical promise of literary hypertext. The idea is not just to establish digital writing practices as one further literary specialization among all the others. The goal of an all-over writing project has not changed since the work of Ted Nelson: to renew literary scholarship as such.

Endnotes

[1] Handling images is still something of a strong-man act, at least in applications that I use in my own writing life – which is non-extreme but I think not unrepresentative, for literary scholars with some investment in e-lit. For example, I went over a year using less than 1% of the capacity on my gmail account, but then the account reached 50% capacity after I circulated among a few friends, resized photos from a single vacation, in a single day.

[2] One instance of "all over" textual distribution is self-exemplifying in the present essay: namely, the sentences leading up to this point in the essay also serve as an introduction to a companion essay, "Electronic Literature as World Literature," under consideration for print publication in a special number of Poetics Today on the topic, "Writing Under Constraint." Otherwise, there is no overlap between that essay and this one.

[3] I have described the "Semantic Web Applicability" to literature in an essay at the Electronic Literature Organization website:
http://eliterature.org/publications/.

A Map of Relations: Three Interfaces in the *Electronic Book Review*

Ewan Branda

Introduction

Over the past fifteen years I have been working both as an information system designer and as a practitioner and teacher of architecture. My role in the *Electronic Book Review* has therefore always been *architectural*. By this, I am referring to two things: first, the design of the principles underlying an information system; and second, a particular way of thinking about that design in which the logic of the interface is seen in structural terms.

There have always been between architecture and informatics strong metaphorical and literal connections. These have, however, tended to be asymmetrical: the information sciences have often used architectural tropes, borrowing terminology (such as "information architecture" or the recently revived archaic verb form of the word "architect" to mean the design of computing systems) and theory (such as Christopher Alexander's design patterns). On the other hand, architecture's use of information technology is for the most part literal. Where architectural tropes in informatics bring with them a degree of abstraction—a translation from practice to principles via metaphor— the incorporation of informatic tools into architecture through the techno-euphoric discourses of the design studio is strictly material. I am therefore concerned in my own research with how such a relationship might be made symmetrical, how information technology might be interpreted in such as way as to provide architectural thinking with new epistemologies. In particular, I'm interested in how one might do so by distancing both "information" *and* "technology" from their almost exclusive association with computing machinery, particularly in Anglo-American discourse.

Spatial information utopias such as the early 20th century Mundaneum of the information scientist Paul Otlet and the later Centre Pompidou in Paris suggest that architecture itself operates as an information technology: it acts as an interface to the information archive; it interoperates within a network of other information systems; it offers a framework within which new information can be produced. As part of an information system, then, the *ebr* interface is a fundamentally architectural problem, which is not simply to say to that it is organized by an "information architect" but rather that it participates in the organization and display of information at a deeper, structural level: it is about mapping relations.

Focusing on their "architectural" aspects, I would like to discuss three types of interface in *ebr*. I will focus on information retrieval problems, specifically those of "relevance" and "aboutness" and how these problems can be located within the logic of the interface's "thick 2-D" space (to borrow a term from the architect Stan Allen (Allen, 2001)). Our current research involves assessing Semantic Web technologies for the next version of *ebr*, and so toward the end of this short paper, I'll share some observations on the promises and limitations of the Semantic Web in achieving our goals.

Interface 1: Between user and archive

The interface between user and database is concerned primarily with the relevance of documents—which set of documents from a database should be displayed to a user for a given condition. Relevance is both a canonical problem in modern information retrieval systems (of which *ebr* is a class) and a central concern of today's debates on bottom-up versus top-down approaches to classification. What has remained unchanged since the foundations of digital information retrieval were laid in the 1960s is the tendency to locate relevance at the level of inherent semantics of documents and queries, whether the "semantic intelligence" used to identify it is modeled as knowledge representation at a high level or simulated statistically using low-level user-generated tags.

Our approach to the problem of relevance is unashamedly editorial, and is exemplified in the Weave page (http://www.electronicbookreview.com/action/ Search?kw=brigham - fig. 1). Following the lead of digital mix and list culture, we shift the emphasis away from inherent document semantics and onto networks of discourse. The editor classifies essays according to primary and secondary threads and assigns to them unstructured, flat metadata tags. At the same time, guest curators are invited to build personal "mixes", which are personal collections of essays on a specific theme. We also maintain a measurement of editorial activity for each essay in the database. The more active an essay, the more relevant it is deemed to be. In this way, an old essay that suddenly enjoys renewed editorial attention (in the form of being newly posted, glossed, or incorporated into a mix) will find itself bubbling to the surface. In determining relevance we thus make no attempt to extract any essential "aboutness" from individual documents; instead, we determine relevance through relations established by editorial assertions interwoven by the reader.

Put simply, our view is that relevance does not inhere in the text of an essay but rather in the relationships among documents across the archive.

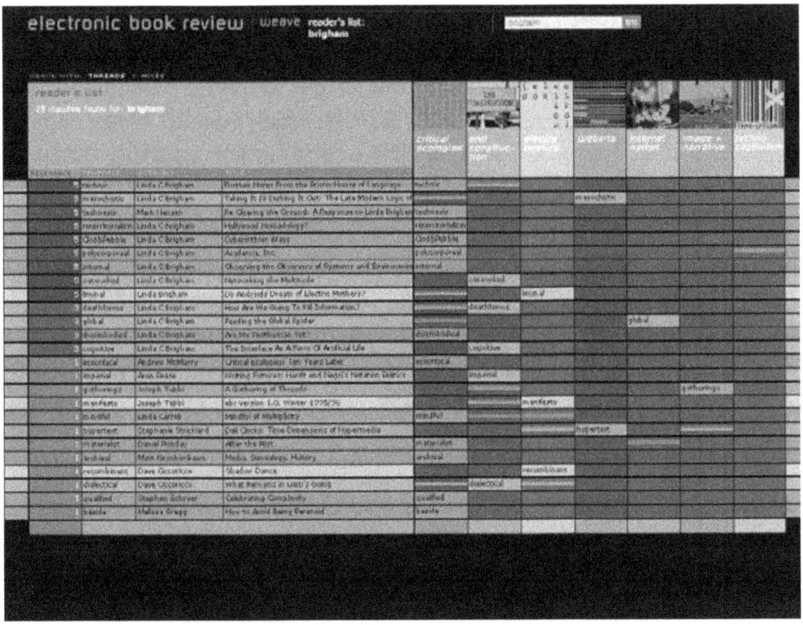

Figure 1: electronicbookreview.com: the weave

A risk in any document map such as the weave is a reliance on indexical modes of display, by which I mean interfaces that passively visualize underlying data. Indexicality is one of the central visual topoi of New Media, perhaps the clearest example being the well-known StarryNight document map (http://rhizome.org/starrynight/) a surface upon which hidden structures of documents and data inscribe emergent patterns, in much the same way that we might perceive the topography of the surface of the ground defined by hidden geological processes. The visual patterns are ontologically dependent on a hidden process, without which there would be simply an empty screen, but at the same time bear a passive relationship to it, a one-way information flow in which the display does not feed back in any way into the underlying process that generated it. The art historian Rosalind Krauss has described indexical signification in artworks as substituting "the registration of sheer, physical presence for the more highly articulated language of aesthetic conventions" (Krauss, 1985). In Starry Night, the reader is witness to a spectacle of information accumulation that, despite the night sky metaphor, is perfectly in keeping with the more general preoccupations with indexicality in New Media.

In the Weave, we extend the logic of the indexical interface launched by such projects as StarryNight, but add to its passive data visualization function a kind of agency: on the one hand, the editors and weavers use this map to build additional mixes and to inform editorial classification of future essays; on the other, the map does not so much display current relations but suggests new connections between texts. By mapping on two axes the interrelated activities of three types of user (editor, guest mixer, and reader—see fig. 2) the *ebr* Weave tries to go beyond the simple registration of the archive's presence to project new configurations. In this way, it is like the kind of map described by the landscape architect James Corner, in that is not simply a passive tracing of a terrain but has "agency" through its production of new territorial configurations (Corner, 1999).

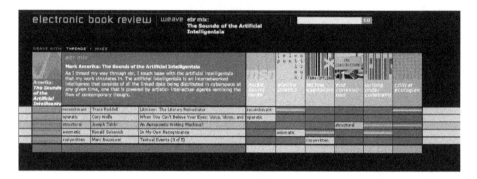

Figure 2: electronicbookreview.com: the weave showing mix

Interface 2: Between user and document

Where the Weave is an interface between a user and the archive as a whole, glossing—the ability to add commentary in the margins of essays—links an individual commentator to a single essay (fig. 3).

From a technical point of view, glossing is simple. We have borrowed the basic architecture of the W3C's Annotea project (http://www.w3.org/2001/Annotea/) in which annotations are stored externally to the annotated document and are linked to specific locations in the target XML using XPointer (fig. 4).

There are a lot of friends of mine who are musicians, some incredible composers who put together symphonies. And in many ways when I was writing this, and *House of Leaves*, there was this musical element.

KB: That's clear. You speak of symphonies – highly formalized, rigid structures, within which you get lyrical movement. This book is so much more constrained by the parameters you set up than, say, *House of Leaves*. There are only so many words per page; there are the symmetries, the increasing and decreasing narratives. The constraints govern the narrative – as in Oulipo. In writing *Only Revolutions*, did you establish a structure before you worked within it, or did the structure and content evolve simultaneously?

MZD: Simultaneously. The thing about Oulipo, as much as I admire the work, it tends to be about the constraint. It's almost like alcoholic writing: it tends to be about the drug, you know. It's deQuincey's opium, it's Kerouac's benzedrine, it's Fitzgerald's booze. They tend to be about that liberating substance, whereas Oulipo tends to be about that constraint.

KB: I used the word "gimmick" once and caught heck from another reviewer.

MZD: I would love to hear your thoughts about what is a gimmick, because it's still something that doesn't make sense to me. I guess if it is for its own purpose – if it's only green so that it is green, and attracts attention, I guess it's gimmicky. I don't really know.

KB: Apart from its structural intricacies, then – what is *Only Revolutions* about?

MZD: At the heart of it are these two kids. They were two kids that I came across. They were impertinent, they were courageous, they were penniless, and most important, they were parentless. They were without anybody. They were sitting on a corner begging for change. And they loved each other. They held onto each other, they looked after each other, they lusted after each other they protected each other. They were each other's world. And the fact that they were without anything was so inspiring. Because they were bold. Maybe they were Homeric gods in disguise. Maybe they truly were Mendicants. But there was something magical about them for me – that absolute attachment to each other. And I looked at them as kind of American Romeo and Juliet. But later I realized that they really weren't – because they didn't have the Capulets and the Montagues out there to separate them, to chase them. There's Bonnie and Clyde, *Natural Born Killers*. They have the law – they have Harvey Keitel – there's someone who pursues them. These two are unpursued. So, you know, for me what became interesting was that ongoing discussion, meditation if you will, exploration about freedom and love. Because freedom is ultimately the quest from anything – to be unrestrained by your circumstances, by your society, by even your own body – whereas love is all about attachment. It's all about the involvement with someone else, which is the opposite of freedom. And yet love allows you to transcend certain things. So, as I was working on this book – and now this is getting back to your question – I realized that I wanted to set them up in a way where they were constrained, they were limited, because their entire quest is how to free themselves.

KB: It's like Sartre's notion of freedom and facticity: there's no such thing as freedom without factical things to be free against.

MZD: Yes, so the word component – the number of words, the structural component – is only part of it. Their freedom is from nature, from society, from

GLOSS
Lori Emerson:
Michael Bovden interviews
Oulipian Harry Mathews on
ebr; here Mathews traces his
own engagement with
constraint to 14th century
and modernist composers.

Figure 3: electronicbookreview.com: essay with gloss

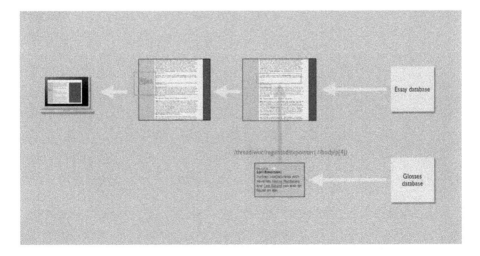

Figure 4: Storing and displaying glosses

More interesting are the questions glossing raises about the annotated text itself. Ideally, a commentator would freely create annotations at arbitrary locations in the text. To design a usable hypertext interface, however, required an a priori segmentation of the text into discrete units; how could glossing operate without predefined units with which the commentator interacts? Moreover, how would a reader navigate an arbitrarily complex nested structure of textual segments, each referenced by an annotation, and still preserve the integrity and flow of the original text? Both of these questions concern breaking down the text into lexias, or units of reading, a well-known structural problem described by Roland Barthes in 1970, and one of the canonical themes of hypertext theory (Barthes, 1974; Landow, 1994). The lexia, according to Barthes, "will include sometimes a few words, sometimes several sentences; it will be a matter of convenience: it will suffice that the lexia be the best possible space in which we can observe meanings".

For lexia, we decided simply to use paragraphs, defined by P elements in the markup (fig. 5). The structured nature of XHTML, along with the fact that *ebr* essays are authored according to strict guidelines, means that this paragraph-based approach poses no particular technical problems in modeling documents. More troubling, however, is the resulting friction between the structural logic of the paragraph and the critical logic of annotation, for, according to Barthes' definition, one can easily imagine a lexia transgressing paragraph boundaries (fig. 6). Clearly, it is problematic to constrain to such a degree the act of interpretation with the paragraph's logic of authorial composition.

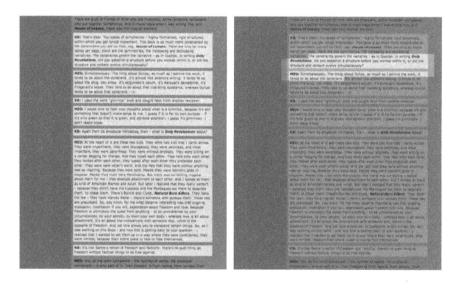

Figure 5: electronicbookreview.com: paragraphs as lexia
Figure 6: electronicbookreview.com: the problem of lexia

Interface 3: Between archive and archive

The final interface that I will discuss involves no human user but only collaborating information systems. We see *ebr* down the road as part of a distributed federation of collaborating web applications, a model now commonplace in enterprise computing but, sadly, rare among the information silos that make up humanities computing. Fig. 7 shows only a hypothetical example, the integration of *ebr* with The University of Virginia NINES project (http://www.nines.org/) and the Electronic Literature Organization Directory (http://directory.eliterature.org). The former project itself argues for this type of collaboration.

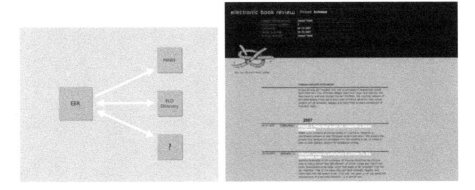

Figure 7: electronicbookreview.com: interaction with other resources
Figure 8: electronicbookreview.com: the thread page

We have taken our first steps into this arena through what we call "enfolded" essays (http://www.electronicbookreview.com/thread/enfolded - fig. 8). These are remote web resources wrapped in the same metadata schema as *ebr* essays, allowing them to be displayed in the *ebr* interface (see, for example, http://www.electronicbookreview.com/thread/enfolded/collaborative). One of the promises made by the Semantic Web, the W3C's proposal for structured metadata, is to solve this type of use case. Unfortunately, even a brief introduction to the Semantic Web is beyond the scope of this paper. I'd nevertheless like to consider it here, both as an architectural pattern for a small set of collaborating applications (as opposed to the universalizing ambitions for which the initiative is rightly criticized) and a protocol for building shared conceptualizations ("ontologies" in Semantic Web language) within a narrow, well-bounded domain. I'd like to do so by considering a more straightforward, almost trivial, use case that is well within reach.

Currently, we store all author names and biographical data in the *ebr* database. At the same time, the Electronic Literature Organization maintains a directory of author biographies with citations of their works. A simple web service could serve this data, making it available to client applications such as *ebr*. Reciprocally, *ebr* could publish a web service, effectively an RSS feed, that would notify subscribing applications such as the Directory about newly published essays. In technical terms, this type of collaboration is simple. Moreover, authority records for data such as author biographies are relatively easy to model and to come to agreement on, compared with more complex ontologies (fig. 9). Yet, even this trivial use case remains unrealized, partially because, in my opinion, justifiable criticisms of the Semantic Web's totalizing ambitions have stifled investigation at a more semantically restricted, local level.[1]

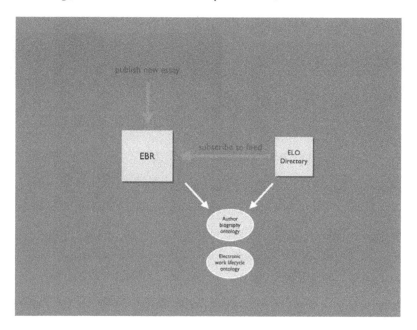

Figure 9: Sharing author data between electronicbookreview.com and ELO Directory

In contrast to the simplicity of the ELO Directory integration, a more ambitious Semantic Web application might allow the automatic generation of mixes. In the spirit of iTunes' "smart playlists", a reader could configure such "Smart mixes" based on parameters such as date of publication, similar *ebr* essays, common keywords, works by a certain author, etc. The application would then monitor a set of Semantic Web-based registered services for new essays to enfold.

Conclusion

This last example raises the question of what should be automated and what should remain under editorial control, a question that has dogged us throughout the project. Even if automation were technically possible, the construction of a universal, semantic ontology for the domain of electronic literature seems unlikely, if not undesirable. Instead, I would like to see ontologies for structural aspects of the archive. At the Experiential Technology Center at UCLA, we have recently completed another project, a database of Ancient Roman architecture, built on simple types of RDF-based reasoning that use ontologies of spatial relations (near, adjacent, facing, etc) combined with ontologies associating ancient authors with topographic locations and defining categories of historic periodization (see http://dlib.etc.ucla.edu/projects/Forum). Similarly, in the domain of electronic literature, common languages for representing document lifecycle and workflow, biographical data, etc. would allow a certain amount of ontological reasoning at the structural level of the archive, allowing for new types of Weaving.

Such local, domain-specific ontologies are more achievable and, happily, more productive than the "world brain" (to borrow from H.G. Wells) promised by the Semantic Web. At the same time editorial control should not be relinquished to what Jaron Lanier, in recently criticizing today's collective online practices, has called "digital Maoism" (http://www.edge.org/3rd_culture/ lanier06/lanier06_index.html). But I would suggest that the folksonomy versus ontology debate is founded on a shared fallacy: both fall prey to the Sisyphean problem of identifying and representing the essential aboutness of documents, ignoring the more accessible and productive meanings residing in patterns of use and conversation.

Endnotes

[1] Joseph Tabbi has recently written on the applicability of Semantic Web technologies to the archiving of literary works (Tabbi, 2007).

Works Cited

Allen, S. (2001). "Mat Urbanism: The Thick 2-D." In H. Sarkis (Ed.), *Le Corbusier's Venice Hospital.* Munich and London: Prestel.

Barthes, R. (1974). *S/Z.* New York: Hill and Wang.

Corner, J. (1999). "The Agency of Mapping." In D.E. Cosgrove (Ed.), *Mappings.* London: Reaktion Books.

Krauss, R.E. (1985). "Notes on the Index: Part 1." In *The originality of the avant-garde and other modernist myths.* Cambridge, Mass.: MIT Press.

Landow, G.P. (1994). *Hyper/text/theory.* Baltimore; London: Johns Hopkins University Press.

Tabbi, J. (2007). "Toward a Semantic Literary Web: Setting a Direction for the Electronic Literature Organization's Directory." Available at http://eliterature.org/pad/slw.html (last accessed February 2008.)

The World Wide Web Evolves*

Formative figures in the creation of the current Web--semantic Web, the grid, and social software--envision Web 3.0.

Dan Connolly
"How the W3C Process Got Its Stripes"

Pat Hayes and Margaret Warren
"Artspeak: The Contemporary Artist Meets the Semantic Web. Creating Formal Semantic Web Ontologies from the Language of Artists"

David de Roure
"Grid of People"

Henry Thompson
"The Humanities, the New Empiricism, and the World Wide Web"

* This session was a round table and thus not conducive to publication in text format. Video archives of the session are available at www.hastac.org.

Racing (through) Domains

Racial attitudes persist in digital media and in race-based surveillance but also in new methods for teaching civil rights history.

Jessie Daniels
"Searching for Dr. King: Teens, Race and Cloaked Websites"

Simone Browne*
"(Im)mobility Documents, Race, and Surveillance"

Irene Chien*
"Orienting Inner Space: Biofeedback Gaming and the Racialized Landscape of Mind, Body, and Spirit"

Michele White
"The Hand Blocks the Screen: A Consideration of the Ways the Interface is Raced"

* Available in digital format in the video archives of the conference at www.hastac.org.

Searching for Dr. King: Teens, Race and Cloaked Websites

Jessie Daniels

INTRODUCTION

For young people who have come of age during the digital era, the notion of "doing research" does not mean going to a library, it means going online to use a search engine to find information (Rheingold 2006). The overwhelming majority of teens born in the U.S. after 1987 are online, with national surveys placing the proportion at 87% (Lenhart, Madden & Hitlin 2005; Roberts, Foehr, & Rideout 2005). Because they have grown up using digital technologies many, though certainly not all, adolescents have a set of nascent Internet literacy and evaluation skills, variously called "digital literacy" (Gilster 1997) or "digital fluency" (Green 2005; 2006; Resnick 2002); yet, many fewer possess an oppositional or critical race consciousness (Blau & Stearns 2003; Taft 2006). The combination of the shift in how young people look for and find information along with the lack of critical race consequences has important consequences for digital media and learning (e.g., see http://digitallearning.macfound.org/), and also for understanding the way young people find and make sense of information about race, racism and civil rights in a digital era.

Race and racism are part of this new digital era in ways both predictable and unexpected. Given the long history of extremist white supremacist activity in the U.S. (Daniels 1997), it is perhaps not surprising that these groups have seized upon the Internet (Daniels 2007; Adams & Roscigno 2005; Whine 1999). What may come as a surprise is that some of these groups have been brought a certain level of sophistication to the amalgamation of white supremacy and the Internet. No longer a battle fought primarily in street protests, struggles over race, racism and civil rights are now being shaped through the use of cloaked websites that challenge the validity of hard-won political victories. In the U.S., many adolescents, like many adults, are naive when it comes to matters of race and racism and this naïveté about racism makes discerning cloaked websites even more difficult. I contend that both digital literacy and critical race consciousness are necessary for understanding race in the digital era. How young people look for information about race, racism and civil rights, and how they make sense of that information once they find it, are the questions to which I now turn.

CLOAKED WEBSITES

Cloaked websites are created and published by individuals or groups that intentionally disguise a hidden political or ideological agenda (Author 2006; 2007; Ray & Marsh 2001). To my knowledge, the term "cloaked" refers to a website that first appeared in Ray & Marsh (2001) in reference to www.martinlutherking.org. I am using the term "cloaked website" in a similar way, and elsewhere (Author 2007) expand the term to include other types of cloaked sites. These can include political sites, such as www.whitehouse.com, which is intended as satire, or websites connected to sexual politics, such as www.teenbreaks.com, which appears to be a reproductive health website, but is, in fact, a showcase for pro-life propaganda.

Here, my focus is on cloaked websites that contain intentionally hidden political and ideological agendas to: 1) undermine civil rights advances of African Americans and other racial/ethnic minority groups and 2) further the ideology of white supremacy. These sites contain virulent anti-Semitism, racism and hate propaganda several page-layers down, or provide links to such information, but the authors conceal both the affiliation to white supremacist groups and the intended political and ideological purpose of this rhetoric. The two cloaked sites under investigation in this study are www.martinlutherking.org and www.AmericanCivilRightsReview.org. The cloaked site www.martinlutherking.org appears at first glance, to be a tribute page to Dr. King, but in fact, is intended to undermine his legacy and further the goals of white supremacy. The site includes text and links to a litany of defamatory information about the civil rights leader, including charges of adultery, plagiarism, and reported connections to communism. First launched in 1995, the site was created by Don Black, an avowed white supremacist with a years long commitment to a racist vision. The site suggests that the entire civil rights movement was a Jewish conspiracy and that the national King Holiday should be repealed because Dr. King was a plagiarist, adulterer and communist. This site also includes links to the work of other white supremacists, such as David Duke, a former Klansman who traded the robes-and-hoods for a suit-and-tie form of white supremacy.

The second cloaked site I address in this research is www.AmericanCivilRightsReview.org, which appears to be a site advocating for civil rights, with images of Che Guevara and Malcolm X on the entry page. However, several page links into the site there is an extensive discussion of the "high self-esteem" of slaves in the U.S. Rather than relying on extremist rhetoric, the site uses a more sophisticated strategy of including links and quotations from oral histories of former slaves recorded by the WPA, to argue that slavery was not a demeaning, violent system and that European immigrants and their (white) descendants suffered hardship on an equal level as that experienced by enslaved African peoples in the U.S. The site is created, owned and operated by Frank Weltner, a member of the National Alliance, a neo-Nazi organization. Weltner also maintains the anti-Semitic website, www.Jewwatch.com. In the aftermath of the Hurricane Katrina disaster,

Weltner also created several cloaked sites, with URLs such as www.InternetDonations.org, to scam money from people interested in helping out the victims. A judge in St. Louis, where Weltner is based, issued a permanent restraining order against the scam websites.[1]

Some writers have suggested that white supremacist groups may use the Internet to recruit new members to join their ranks (Back, Keith & Solomos 1996; Mock 2000), with young children and teens seen as particular at risk for targeted recruitment (Lamberg 2001). While it is certainly the case that extremist, paramilitary racists are using Internet technology for "command, control and communication" (Whine 2000) and this is cause for concern, there is little empirical data to support the claim that unsuspecting web users are unknowingly lured into extremist group membership via the Internet (Author 2007; Ray and Marsh 2001). A standard response to this perceived threat, especially on the part of parents of younger Internet users, has been the development and use of "hate filters," software programs designed to "filter" hate sites encountered through search engines. While this type of software may filter some extremist racist sites (at least English-language sites), these are not the type of white supremacist site that most web users are likely to encounter. In my view, the casual web user is a much more likely possibility for inadvertently encountering white supremacists online via cloaked websites which appear in search results when looking for information about race, racism, civil rights or civil rights leaders. This leads me to the research questions under investigation here: 1) how do adolescents (age 15-19) use search engines to find information about civil rights on the Internet; 2) how do adolescents evaluate information they find about civil rights on the Internet; and, 3) how do adolescents evaluate *cloaked* white supremacist websites they encounter on the Internet?

THE CHALLENGES OF STUDYING RACE & DIGITAL MEDIA

The Internet presents new challenges for qualitative sociological researchers, as a number of scholars have noted in the first decade of online research (Markham 1998; Jones 1999; Wellman 2004). Digital media, such as the Internet, is particularly challenging for sociologists interested in race. In the early 1990s, the web's nascent popularity and the putatively disembodied quality of online interaction caused many observers to remark on the potential of the Internet as a deracinated medium, perhaps most famously in a popular television commercial that touted, "Here, there is no race. There is no gender. There is only mind" (Everett 2002). However, rather than a place where people eschew racial identity a growing body of research demonstrates the ways that people bring their embodied selves to the keyboard and seek out the Internet specifically to reinforce and reaffirm racial identity (Byrne 2007; Kolko, Nakamura, & Rodman 2000; Nakamura 2002).

A leading example of the way that people seek out and use the Internet to reaffirm racial identity is the early emergence of extremist white supremacist individuals and groups online, and a brief review of some of this research illustrates some of the challenges of studying race and digital media. Examining the text of web pages, discussion forums, and newsgroups is the most straightforward method and it is also the most common way of studying the white supremacists online (e.g., Adams & Rosigno 2005; Atton 2006; Back, Keith & Solomos 1996; Bostdorff 2004; Gerstenfeld, Grant & Chiang 2003; Kaplan, Weinberg & Oleson 2003; Levin 2002). More difficult, and less prevalent, are investigations into the connections between online interaction and face-to-face social networks among extremists (Burris & Strahm 2000; Hara & Estrada 2003; Tateo 2005). Most vexing still and least common are studies of the "web user." In other media, this type of research is called "audience reception," and explores how the listener, viewer, or reader interprets the "text," whether that text is visual (as in films or television shows) or printed (as in novels or newspaper articles). Sonia Livingtone (2004) has suggested that the terms "audience" and "reception" do not work well for digital media for a variety of reasons, such as interactivity (rather than one-to-many, with producer and receiver separate as in broadcast media). When it comes to empirical explorations of how people find, read, and interpret extremist rhetoric on racist websites, there is scant research. An important exception to this is the work of Lee & Leets who examine how adolescents respond to what they call "persuasive storytelling" online by hate groups (Lee & Leets 2002). Lee & Leets found only minimal effects on adolescents who were infrequently exposed to explicit hate messages. However, their research did not explore how adolescents might be exposed to these messages; and, it only focused on explicitly racist sites, and not on cloaked websites. In this paper, I address this gap in the emerging body of knowledge about race and the Internet, and specifically I address the question of how teens find information online about race and ask how they interpret cloaked websites.

METHODS

This is a pilot study and is therefore meant to be exploratory and suggestive rather than exhaustive or definitive. I conducted the study in January and February 2006 and asked adolescents (ages 15-19) to use the Internet to search for information and to evaluate two pre-selected pairs of websites about Dr. King and about the civil rights movement.

STUDY DESIGN

I utilized a mixed-method study design, which included search scenarios, paired website evaluations, and a technique known as "talk aloud" (also referred to as "think aloud"). There were two search scenarios. The first included asking participants to "find information on Martin Luther King as if you had a report to write for school." The second scenario asked participants to "find information about the goals of the civil rights movement as if you had a report to write for school." As they reviewed the results of their query returned by the search engine, I asked them questions about what they saw, what looked interesting to them and why, and which websites they would select to read.

After completing the search scenarios task, I asked the participants to evaluate the differences between pairs of websites. The first pair included the legitimate King Center site (www.thekingcenter.org) and the *cloaked* Martin Luther King (www.martinlutherking.org) site; the second pair included the *cloaked* American Civil Rights Review (www.americancivilrightsreview.com) site and the legitimate Voices of Civil Rights (www.voicesofcivilrights.org) site. I pre-selected these sites based on the similarity of content and traffic. For example, the traffic in 2006 to the websites for the King Center (indicated by the blue line) and the cloaked Martin Luther King site (indicated by the red line) are nearly identical, with an overall peak in February, which is African American History Month (Figure 1).[2]

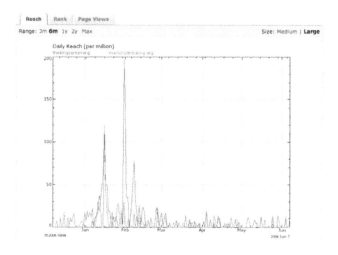

Figure 1: 2006 traffic to the websites for the King Center (indicated by the blue line) and the cloaked Martin Luther King site (indicated by the red line).

I minimized the windows for all four websites on the computer, and introduced them to the participants in pairs. I made sure to change the sequence, introducing a cloaked one first, followed by a legitimate site, and then reversing the order. Some participants had already found these sites during the initial search scenario, and for them, I asked them to look at the sites again, in relation to the paired website and talk aloud about which site they would choose as a source of information if they were forced to select one for a school report. In the paired website evaluations, I pre-selected two pairs of websites. In each pair, one website was a legitimate source of information about Dr. King or the civil rights movement, and the other was a cloaked site about the same subject.

During both tasks, the search scenarios and the paired website evaluations, I asked participants to "talk aloud" about what they were doing. The "talk aloud" technique, which is common in usability studies of graphic user interface (GUI) website design and frequently used by marketing firms, asks web users to describe what they are doing, seeing, thinking, reading, and clicking on, -- and why they are making those choices -- as they navigate a web site (van Someren, Barnard and Sandberg 1994). Completing both tasks took participants approximately thirty to forty-five minutes. I recorded these sessions using a digital video camera, recording audio of the participants' voices, and accounts of their searching and evaluating the web, and capturing video images of the computer screen as they searched.

ANALYSIS

I transcribed the audio portion of the interviews and noted in the transcripts what was on the computer screen at the same time so I could recall which websites the participants were referring in their interviews. I also noted the sequence of their navigation through the sites, the images on the screen, and the way they responded to these. I then coded the transcripts by theme, and analyzed them for similar and discordant themes across interviews, and for consistencies or changes in patterns within interviews. This process, although time-consuming, is useful because it situates the web user in relation to the visual images, the text and hypertext of the web. Reviewing the video portion of the interviews, and noting it in the transcripts, also provided additional information about the way participants searched, navigated, read, and made meaning of search results or of a particular website.

THE SAMPLE

This research utilizes a convenience method of sampling and includes a sample size of ten (N=10). Participants for the study were recruited through a variety of means, including through a youth-focused human rights foundation, word-of-mouth, printed flyers and online bulletin board postings. The majority (N=8) were recruited from the online bulletin board, one through word-of-mouth, and one from the foundation. Almost all (N=9) were female, and came from a variety of racial/ethnic backgrounds (1 African American, 1 Asian-Chinese, 2 White, 2 Latina, and 3 South Asian); the one male respondent was Latino. All indicated that they were born in the U.S., and all were enrolled in high school, in the 11th or 12th grade, at the time of the study. Participants under age 18 who participated in the study were required to get parental consent and were guided through the informed assent process. Participants 18 and over were guided through the informed consent process. Except for the participant at the foundation, all participants were asked to travel to my faculty office on a college campus in the city, to complete an interview that lasted less than an hour. Participants usually arrived alone to the interview, although one participant brought her mother, who sat quietly while we completed the interview. Participants who completed the interview received a $20 stipend for their time, and were given information about Internet searching during the debriefing following the interview.

Given that a majority of the participants volunteered for the study via the online bulletin board postings, it is likely that this is a sample of relatively digitally fluent and Internet-savvy teens. Of course, because of the convenience of the sampling strategy employed these results are not generalizable to all teens, or even all teens using the Internet in New York City. However, the Pew Internet & American Life Project has conducted large, national, random sample survey research into the online practices of adolescents. In 2005, Pew researchers found that 87% of adolescents aged 12-17 used the Internet and 51% use the Internet on a daily basis, and 76% get news or information about current events. This is in contrast to adults, who are less likely to use the Internet, with 66% of adults using the Internet (Lenhart, Madden & Hitlin, 2005). This research also indicates that among older teens (15-17) girls are "power users" of the Internet and search for information about a variety of subject areas; and, they are more likely to use a variety of digital technology, including email, instant messaging, and text messaging, than their peers (Lenhart, Madden & Hitlin, 2005:6). It is likely that the sample for this study includes participants who are similar in their web usage to the national sample. In particular, the fact that I was able to recruit a majority female sample using an online bulletin board posting suggests that these young women are typical of the "power users" identified in the Pew research.

FINDINGS

Results of this research indicate that searching for, finding, and evaluating information about race online is a complex process with many components. Here, I have tried to lay out that process in as much detail as possible. In the transcripts included the image and sites on the computer screen are noted in brackets and italics; the interviewer questions are in plain text; and, the participants responses are noted in bold text.

SEARCHING FOR DR. KING & FINDING DAVID DUKE

In the search scenarios, I asked participants to use any search engine and any search terms to find information about Dr. Martin Luther King, Jr. and then asked them to do the same to find information about the goals of the civil rights movement, "as if doing a report for school." Nine of the ten participants selected the search engine Google (the remaining participant used the search engine Yahoo.com). All of the participants used the same search engine throughout their interview, and did not change or switch to another search engine. The participants also used similar search terms. The most commonly used search terms for the first scenario were: *"martin luther king"* or *"martin luther king + biography."* And, for the second scenario, the most commonly used search terms were: *"civil rights," "civil rights movement"* and *"civil rights goals."*

When asked about how they evaluate the search engine results, most said that they relied on the order that search results appeared as a valid and reliable way to evaluate whether or not a site was trustworthy. This was a consistent theme across the interviews and is reflected in this quote from a participant reported that they would "never" go beyond the first page of results in their research of a topic, as did this participant:

I actually have never, I think, in my life gone to like the third page, or the second page, because I just stop at the first page. ... because I mean, there must be a reason why everything's on the first page and the rest of the stuff is later. (age 16, white)

In a sense, this young woman is correct when she says, *"there must be a reason"* for the results on the first page. There is a reason and it is an algorithm. Given the huge popularity of Google as the search engine of choice by so many, we might expect that there would be wide familiarity with how the search engine works. As it turns out, this is not the case. Actually, different search engines work differently, and the way Google works is through a fairly complex algorithm that includes a web crawling robot, the Google indexer, and a query processor. PageRank is Google's mechanism for ranking one web page higher than another. Central to this mechanism are links from outside pages; each link from an outside page to a website is, in Google's evaluation schema, a vote for the "importance" of that site (Sherman and Price, 2001). So, while there is a reason that those results appear on the first page, it is not because someone sitting in an office at Google headquarters has read and evaluated each site and rank ordered them based on an agreed upon set of criteria. In fact, because of the way Google's algorithm works, it is possible to intentionally manipulate the ranking of a site by linking to a page using consistent anchor text. This is commonly referred to as "Google bombing" and has been used a number of times as a form of political critique of the Bush administration; thus, because people on a number of websites across the Internet have repeatedly used the same linking anchor text, now anyone can type the search terms "miserable failure" into Google and get the first result to be a link to the "Biography of George W. Bush" (Byrne, 2004; Kahn & Kellner, 2004). When I asked the participants if they had ever heard of a "Google bomb," not one said that they had and were perplexed and amused when I showed them the "George W. Bush" results for the "miserable failure" search. Trusting the results on the first page of Google might not be an issue for understanding race except for two key points: 1) when searching for information on race, racism and civil rights, cloaked white supremacist sites appear alongside results for legitimate sites; and 2) people, like the young woman quoted above, implicitly trust the order of results as a valid and reliable mechanism for assessing trustworthiness.

The cloaked site www.martinlutherking.org consistently appears third or fourth on the first page of results in Google when using the search terms "martin luther king," and this, along with the URL, has implications for how young people find information about race, racism and civil rights. Typical of the way participants in this research evaluated the cloaked site when it appeared in search engine results was this young woman:

[Computer screen: Opens Google, uses search terms "martin luther king" without quotations. Once the search results are returned, she scrolls the page quickly, using

the mouse button.]
Right now, I'm just reading the sites, to see what they're about, to see which ones are easier for me.

Ok, and what kinds of information do you look at? What pops out at you?

I guess maybe something like this would pop out, an article from The Seattle Times....about his life, and impact.

Ok. And is that a link that you might click on?

I would just look at it, I wouldn't click on it yet... but this one....
[Computer screen: Points her mouse to the martinlutherking.org link returned third in the list of results from Google.]
... this one looks good.

You think you would click on that one?

Yeah, because the site itself, it says, "Martin Luther King dot org" so I guess they're dedicated to that. (age 18, Latina)

Here, in the span of just a few seconds after typing in the search terms looking for information about Dr. King, this young woman has come across the cloaked white supremacist site and evaluates it positively, along with a legitimate site about Dr. King hosted by *The Seattle Times* newspaper. In part, this participant is responding to the anchor of the site's universal resource locator (URL); in other words, the fact that all of the web address is made up of the civil rights leader's name makes it seem legitimate. She is also responding to the suffix, or ending, of the web address: "dot org." This kind of response to the URL www.martinlutherking.org was a consistent theme throughout the interviews. Participants understood the suffix .org to mean that a site was a legitimate source of information, as this young woman explains:

[Computer screen: Scrolls up and down the list of search results, including the martinlutherking.org link returned third in the list of results from Google.]
Ok. Anything else about the URL that lets you know it's trustworthy?

That's about it. Basically, like the source where it's coming from. I mean, if it's like a personal web page or something, they just have information about him, I wouldn't go there.

And how do you know when it's a personal web page? How can you tell?

Well…. Ok, like if it's dot edu, then you know it has to do with education, like a university or something. And, you know, like this one…

[Computer screen: Scrolls over the results for a legitimate site, hosted at Lucid Café, a web portal created and run by Robin Chew a web developer and marketing executive in San Francisco.]

Lucid Café, that doesn't look too, I don't know… the title looks serious, but the URL…

Alright, based on the URL you wouldn't go there even though the title and the description look ok.

Yeah. And, dot org, too….

Yeah, and what does that mean for you?

I don't know what it means, actually…. [laughs] … organization?

No, it's fine… I don't mean what does it actually mean, I meant, what does it indicate to you…?

Oh, ok… again, it's more of a trustworthy website. Because, dot coms are everywhere, and dot org and dot edu are more specific. (age 18, African American)

The fact that URL's ending in "dot com" are more common ("everywhere"), leads this participant to conclude that the less common .org websites are more trustworthy. For the most part, Internet literacy skills-based classes have instructed web users to "read the URL" as a first step for evaluating the legitimacy of a website, and to "trust" URLs ending in ".org" more than those ending in ".com." Thus, this participant is doing precisely as she has been taught. While it is possible to read the URL of a site and sometimes ascertain where the site is hosted or who is sponsoring it, it is also possible to disguise a site by means of a clever or nefarious domain name registration. In the case of the martinlutherking.org site, the cloaking of white supremacist political and ideological goals began when Don Black registered the domain name and launched the site in 1995. This suggests that racial politics in the digital era have shifted to a new location in which domain name registration is a site of struggle over racial meaning. It also suggests that typical approaches to teaching Internet literacy skills are inadequate on their own to meet the demands of this new form of struggle over meaning in racial politics. Also necessary is a basic understanding of racism and the struggles against it; without at least a basic understanding of this, the possibility of being duped by a cloaked white supremacist site is much greater.

The lack of understanding about racism and the civil rights struggle can contribute to an inability to recognize a cloaked site, and is illustrated in the following interview. This is an account of the first four and a half minutes from the time the participant begins the search scenario to look for information about Dr. King (this interview conducted by a research assistant):

[Computer Screen: Google results for search terms "martin luther king"]
And see what results come up? Am I looking for any particular website?

No, just any website that comes up, maybe the first three.

Ok, and I'm finding information on his life history?

[Computer Screen: Stanford University's site about Dr. King]
Ok, this is a website by Stanford University so I think it would be pretty well established and accurate. His biography is on here. Other sites that we found included.....
[Computer Screen: back to Google results]
...the Seattle Times on Martin Luther King on the Civil Rights Movement, and...
[Computer Screen: Clicks on the Seattle Times page...then, back to Google results]

Could you try clicking on some of those links and see what they say also, like once you get inside of the website?

[Computer screen: Clicks on the www.martinlutherking.org...main page]
Sure....there's a Martin Luther King pop-quiz, there's some historical writings, essays, sermons, speeches, ...
[Computer screen: Clicks on what seems to be a broken link on www.martinlutherking.org, to 'Historical Writings' the link takes more than a couple of seconds to load, and she abandons it as a broken link. Then, she skips to the 'Truth about King' link without comment, and clicks on 'Death of the Dream' link, subtitle, 'The Day King was Shot']
...there's information here about the day he was shot. It has some photographs of Dr. King the day before his death...and some information about what happened the night before he died which is not apparently public knowledge, or, yeah, it's not like common knowledge....
[Computer screen: Goes back to scrolling over the links on the right. Clicks on suggested books and on that page several titles appear, including, "Plagiarism and the Culture War," and a picture of David Duke's on the cover of his book "My Awakening."]
...then, there's some information on some books that were

written by Dr. King and biographies that were written by, about him by other people…..
[Computer screen: Clicks on "Truth about King" page]

Do you know who that person was? *(referring to David Duke)*

No, I have no idea.
[Computer screen: Clicks back to the "Suggested Books" page]

David Duke, have you ever heard of him?

No.

You've never heard of him?

Uh-uh (no). Who is David Duke?

He's a Klan leader.

Oh, is he? I had no idea. I actually don't know much about the civil rights movement at all.
[Computer screen: Reading Duke book description more closely now.]
Hmmm…. Interesting. It's interesting how that would be on Martin Luther King's website. (age 18, South Asian)

In less than five minutes from when she began using a search engine to look for information about Dr. King, this young woman has selected a cloaked white supremacist site and is reading a page that contains the views of David Duke, an avowed white supremacist, yet she does not recognize that this site is cloaked. Consistent with conventional Internet-literacy skills training, she is reading the URL as legitimate. What is lacking here is not her Internet-literacy skills, it is her understanding of the historical context of racism in the U.S. and David Duke's place in it. As she says, "I actually don't know much about the civil rights movement at all." Although it is may be possible to have an understanding of racism and the civil rights struggle against it in the U.S. and still not know who David Duke is, not knowing seems to suggest a lack of critical awareness about contemporary racial politics. This young woman is certainly not alone in this lack of critical awareness and it is not surprising given the push toward a mediocre testing-based educational system that lacks critical thinking in general (Aronowitz & Giroux, 1993), and is completely lacking any analysis of racism, either historical or contemporary (Feagin, 2006).

EVALUATING CIVIL RIGHTS ONLINE: PHOTOGRAPHIC EVIDENCE AND VISUAL CUES

When asked to evaluate the differences between the legitimate site and the cloaked site, participants used a variety of strategies to assess the differences between the pairs, including digital photographs and visual cues. The web is a visual, as well as textual, medium (Smith & Chang, 1997). As such, those who have grown up using the web expect to find visual and photographic images in their search results. Indeed, they rely on these as important sources of information, as this young man explains while exploring the (legitimate) Voices of Civil Rights web site:

> **This site looks good, I mean, it has a lot of pictures and photos so you can see for yourself what happened. (age 18, Latino)**

Seeing photos as a window with a view of "what happened" was a consistent theme across the interviews. Here, another participant describes her initial impression of the cloaked Martin Luther King site:

> **First thing I notice is the colors...and although the colors are more, are duller, they're in black and white. And, his picture, the picture of Martin Luther King that makes a major difference. Because, you know, it's this picture that attracts all your attention to it. (age 17, Asian-Chinese)**

And, a third participant describes her impression of the *Seattle Times'* use of photographic images this way:

> **Well, they have a photo gallery which I would probably click on because photos are, photography interests me, so...**

Ok, and would that be useful to you in doing your report on King, and if so how?

> *[S: Clicks on a black-and-white photograph of Mrs. King kissing a smiling Dr. King, there is a caption to the right]*
> **Well, like this photo, without even reading the caption...I already know what he looks like so I know that's him, that's his wife and it looks like a good occasion. (age 16, South Asian)**

Visual images are not simply "decoration" for a site but carry messages, convey meaning, and suggestion connotations for these participants. This expectation of, and reliance on, visual images was consistent across all the interviews. Without visual images a particular website was not only deemed less reliable, it was simply less interesting, and often discarded as a possible resource, as this participant describes her assessment of a site that was text-only with no visual images:

> **This site seems awfully wordy...I don't know that I would use**

this one. (age 15, White)

In particular, visual images that appear to be historical photographic images, were a significant part of what the young people in this study were looking for and expected to find when they went online to search for information about Dr. King or the civil rights movement. And, photographic images seemed to carry the weight of authenticity for them, because they reportedly allow one to "see what happened." This reliance on the supposed veracity of photographic images is ironic at a time that some have referred to as the "post-photographic era" (Mitchell, 1992). In the digital era, the widespread use of software that can alter photographs in ways that are virtually imperceptible to the untrained eye makes photography less a "window" on the truth and more of an act of interpretation. That this has significance for racial politics became evident in 1994 when O.J. Simpson was arrested and a photograph of him appeared on the cover of TIME magazine in which the color of his skin was darkened in the photograph.

Aside from photography, other visual cues a major way teens reported that they evaluate civil rights information online. Background and text color, font, layout, and the entire graphic user interface (GUI) of websites were primary criteria used to evaluate whether or not a site was trustworthy, as this participant describes her assessment of one of the cloaked sites:

> **This site looks like someone, you know, just an individual created it. It doesn't look very professional. (age 17, South Asian)**

Here, a site that does not look "professional" is deemed an untrustworthy source of information. Conversely, a site that has a GUI that gets positively evaluated is deemed to have trustworthy content. The distinction between a site that is "professionally" designed and one that "an individual" created is the important distinction here, as this participant illustrates in her evaluation of the paired websites:

> **[Computer screen: Clicks on the cloaked MLK site …]**
> **This one certainly looks less professional.**

And what tells you that it looks less professional?

> **Uhm, it doesn't have a clean lay out, like this one…**
> *[Computer screen: Clicks back to the King Center…]*

Ok, and so…what does that mean? What do you believe about the site or the people who created it?

> **Well, this one was designed, like they hired someone to design**

it… (age 16, South Asian)

In these examples, both the participants take visual cues from graphic design about the trustworthiness of the information contained there. While visual cues are important elements in evaluating web content, they can also be easily manipulated. If the cloaked websites under investigation here made use of "more professional" web design graphics and layout, it would make them much more difficult for these young people to distinguish them from legitimate sites.

CRITICAL RACE CONSCIOUSNESS & ASSESSING 'BIAS' ONLINE

Thinking critically about race is crucial to being able to distinguish cloaked websites from legitimate civil rights websites because this is, ultimately, a political distinction. Without the ability to think critically, all websites are reduced to the level of personal opinion without reference to the power relations that imbue racial politics. And, without a critical race consciousness, one website is just as "legitimate" or "biased" as another. A number of the young people in the study evaluated websites in a way that reflected a lack of critical race consciousness, and it made evaluating the sites more difficult:

[Computer screen: Clicks from cloaked site to the King Center site]

Well, you know, in looking at this site, it appears to be created by his widow, or his family, so, it could be biased. (age 17, Latina)

In this instance, the legitimate civil rights website sponsored by the King Center is evaluated as a less than reliable source of information because it is affiliated with Mrs. King, and therefore, "biased." This young woman is doing what she has been taught in skills-based approaches to Internet-literacy, to "look for bias." Yet, in this instance, it leads to the erroneous conclusion that the King Center site might not be a good source of information about civil rights or Dr. King. While the King Center site certainly presents information from a point-of-view, it is precisely this point-of-view -- situated in the struggle for civil rights and against racism -- that gives it credibility. Another teen assesses "bias" in this cloaked site:

Do you know who published this site, who's behind it?
[Computer screen: Looking at graphic on the top of "High Self Esteem for Many American Slaves" page on American Civil Rights Review]
Uhm, Currier & Ives?

No.
[Computer screen: Spends some time clicking through the site… then comes to a page that has all those badges, etc. on it, and a 'copyright' link to copyright-language on a page hosted by Cornell University.]

Is it by Cornell?

No.

[Computer screen: Reading from the text on the page about slavery, American Civil Rights Review]

I mean, I don't think I would disagree with it. I'm sure there are some slaves that were treated well. So, I can understand their point of view. There's always two sides to everything. (age 17, South Asian)

In this case, the young woman assesses that this site, as just another "point of view," another "side" on a two-sided argument. She is also unable to ascertain who it is that's publishing the site, which is hosted by anti-Semite and racist Frank Weltner who is advocating on this page for a re-writing of the history such that plantations were "sanitary, humane and relaxed," workplaces rather than institutions predicated on human misery. As in the previous example, this illustrates how a lack of critical thinking about racial politics offline can lead to misreading online.

For young people who possess critical race consciousness, recognizing cloaked websites is within their reach. The following is how another teen approached the same, cloaked website created by Frank Weltner:

So, I'm looking at the URL and it says, American Civil Rights Review, slash, slavery, so I'm looking at the main thing, it says American Civil Rights, so it's probably something that I would depend on. And, now I'm looking at the picture of a cotton plantation on the Mississippi River, and you know, plantations and slaves are related a lot, so that relates to slaves. I'm going to just scroll down.... there does seem to be useful information.
[Computer screen: reading from the American Civil Rights Review]
"Idyllic View of American Slavery" they just have pictures, I would rather have, oh, they're actually talking about how the artist basically portrayed the slaves.
[Computer screen: reading from the American Civil Rights Review]
"Now notice how the artist has painted the slaves in relaxed positions."
[Computer screen: pause – reads silently]
It kinda sounds like, like I'm reading this, "were the slaves mistreated" - it says "sometimes" ...
[Computer screen: points to screen]
...and that just throws me off, because I think, yes, slaves were mistreated all the time. And, then it says, "sometimes."

And so what does that mean? What do you think now that you've read that?

Now I don't think it's accurate anymore! Because it says...
[Computer screen: reading from the American Civil Rights Review]
..."sometimes but most probably no more than were other workers including whites." I highly would disagree with that, it sounds so false to me because most of the slaves, they were all black. And white people would not have been treated the same way. And, then it goes into [reading] "Europeans were sometimes given the hardest jobs," when you're talking about slaves and then they're going to Europeans which were obviously not treated the same as slaves because the slaves weren't even treated like people. So, that just throws off everything.

So now what do you do with this site? You said before that the URL looked good and it might have some useful information.

I wouldn't use it.

You wouldn't use it?

No, because even if I find other information that seems accurate, this just makes the whole thing biased to me. Because, to me, the answer would be "yes" there's no "sometimes" or "no," it's "yes." So, I wouldn't even use this. (age 18, South Asian)

Here, the participant decides not to use the cloaked site based not because of an evaluation based on her Internet-literacy skills, but rather on her ability to think critically about race. She reads the text about slaves being mistreated "sometimes" and says, "that just throws me off." Ultimately, she decides the site is not a credible source of information and she would not use it. And, even with her negative evaluation of this site, she uses the same language as the previous two interviews, when she says that the site is "biased," simply back into the skills-based language of Internet-literacy curricula. New ways of thinking about racism in the digital era will have to move beyond two dimensional notions of "bias" in which there are "two sides to everything."

DISCUSSION

Findings from this research suggest that even Internet-savvy teens have difficulty distinguishing between legitimate civil rights websites and cloaked white supremacist websites. While adolescents who have grown up with digital media are fluent in some aspects of the use of technology, they often lack skills in critical thinking and critical race consciousness which would enable them to recognize cloaked websites and distinguish them from legitimate civil rights websites. However, these findings are not generalizable to all teens due to the convenience sampling strategy and the small sample size. Future research should include a randomly selected sample and larger sample size. In addition, teen girls were overrepresented in this sample and more research is needed on teen boys and their use of the Internet.

For the sociologists interested in race, the findings of this research suggest that it is important to understand digital media and the ways this is becoming a new, contested terrain of meaning about race, racism and civil rights. This research has implications for those sociologists who are in the classroom, because it is increasingly likely that students will rely on the Internet as a resource for their information about race, racism, and civil rights. For researchers primarily interested in Internet technology and digital literacy, these findings suggest that it is important to move beyond a skills-based approach and to think critically about race and racism.

These two arenas – digital media and race – come together in the new Internet era and in ways that were not anticipated in commercials that claimed "here, there is no race." Instead, race and digital media are changing both the ways that we think about race and the ways that we think about the Internet. For example, the URL is now a racially contested terrain. As for cloaked websites, they shift and expand the struggle over racial politics to domain name registration and GUI. The decision to register the domain name "martinlutherking.org" in the early 1990s was a prescient and opportune move for advocates of white supremacy; failure to do likewise was a lost opportunity for advocates of civil rights. Recognizing that domain name registration is now a political battleground, a number of civil rights organizations have begun to reserve domain names to prevent hem from being used by opponents of racial justice. The NAACP has registered six domain names that include the word "nigger" and the ADL registered a similar number of domain names with the word "kike" (Festa, 2002). However, registering offensive epithets is only one small part of the struggle. The move by racist opponents of civil rights to register the esteemed symbols of civil rights as domain names, such as Martin Luther King, and use them to undermine racial justice is one that was clearly unanticipated by civil rights organizations. To be effective, cloaked domain names such as www.martinlutherking.org or www.AmericanCivilRights Review.org, rely on the naïveté of their target audience, particularly white people. The vulnerability of these cloaked sites however lies in their inexpert GUI and rudimentary designs, which makes them easier to spot. The problem is that poor graphic design and web layout are technical bugs that are easy enough to fix. Once the elements are resolved, reliance on these visual cues will not be enough to distinguish cloaked sites from legitimate ones. Instead, it will be a much more difficult task in which people will need to parse the rhetorics of white supremacist ideology and progressive racial politics based on the overall *content* of the site, rather than merely the *color* of its graphics.

Obviously, unsuspecting white people are not the only ones that read these cloaked sites, people of color, particularly youth of color, also read these sites. For young people of color, reading cloaked sites means having their own culture and history distorted in the re-telling, and this is characteristic of the epistemology of white supremacy. This, however, is not new or unique to digital media. For people of color have had their culture and history distorted by whites, both those with and without good intentions for many centuries. Black feminist epistemology, suggested by Patricia Hill Collins and others, may hold some keys for understanding these sites. Collins's epistemological stance places an emphasis on lived experience as a criterion of meaning and suggests that ideas cannot be divorced from the individuals who create and share them (Collins 1990). This is where youth of color who have experience racism may have an advantage in critically evaluating these sites. If they draw on lived experience of everyday racism and do the critical work of evaluating *which* individuals are creating the ideas contained in cloaked websites, then they may have an advantage over those steeped in the epistemology of white supremacy that reinforces illiteracy about racism (Mills, 1997).

CONCLUSION

Sociologists, particularly those interested engaged in qualitative research about race and racism, must take into account digital media. Digital media is neither a panacea for eliminating racial inequality, nor is it a dangerous lure for young people who can be duped into joining hate groups. A more nuanced understanding of both race and digital media suggests that the new racism online looks, in many ways, like the old racism, and our culture and institutions are steeped in it. Within the U.S., the culture and institutions were originally formed by slave-owning elites (e.g., Thomas Jefferson's *Notes on the State of Virginia*) and this legacy of white supremacy endures. Young people, depending on their lived experience offline, may use digital media to resist or to reinscribe white supremacy, and engaged adults can influence which of these paths they choose. Trying to understand cloaked websites exclusively in terms of a skills-based Internet literacy which lacks critical thinking about race and racism is doomed to fail. The emergence of cloaked websites calls for different kinds of literacies: a literacy of digital media, to be sure, and new literacies not merely of "tolerance," but also literacies of social justice that offer a depth of understanding about race, racism and multiple, intersecting forms of oppression. At stake in this shifting digital terrain is our vision for racial and social justice.

Endnotes

[1] There is no similar injunction against www.JewWatch.com. In 2004 there was a grassroots effort to convince Google to remove the site from its search engine, but these efforts failed.

[2] These charts are from the web traffic site Alexa. There is slightly higher traffic recorded for the King Center site at the end of January, 2006 around the time of the death of Mrs. Coretta Scott King.

All research and subsequent modifications were approved by the Hunter College, City University of New York, Institutional Review Board (Protocol # HC-080513561).

REFERENCES

Adams, Josh and Vincent J. Roscigno. 2005. White supremacists, oppositional culture and the world wide web. *Social Forces* 84:759-778.

Atton, Chris. 2006. Far-right media on the Internet: culture, discourse, and power. *New Media and Society* 8:573-587.

Aronowitz, Stanley and Henry Giroux. 1993. *Education still under siege.* Westport, CT: Praeger/Greenwood.

Back, Les, Michael Keith and John Solomos. 1996. The new modalities of racist culture: technology, race and neo-facism in the digital age. *Patterns of Prejudice* 30:3-28.

Blau, Judith and Elizabeth Stearns. 2003. *Race in the schools: perpetuating white dominance?* Boulder, CO: Lynne Rienner Publishers.

Bostdorff, Denise M. 2004. The Internet rhetoric of the Ku Klux Klan: a case study of community building run amok. *Communication Studies* 55:340-361.

Burris, V., E. Smith, and A. Strahm. 2000. White supremacist networks on the Internet. *Sociological Focus* 33:215-234.

Byrne, Dara. 2007. The future of (the) race: identity and the rise of computer-mediated public spheres. In *Race and Ethnicity*, edited by Anna Everett, MacArthur Foundation Series on Digital Media & Learning.

Byrne, Seamus. 2004. Stop worrying and love the Google bomb. *Fibre Culture.* 3 http://www.journal.fibreculture.org/issue3/issue3_byrne.html (Accessed February 28, 2007).

Campbell, Alex. 2006. The search for authenticity: an exploration of an online skinhead newsgroup. *New Media and Society* 8:269-294.

Castells, Manuel. 2001. *Internet galaxy: reflections on the internet, business and society.* Oxford: Oxford University Press.

Collins, Patricia Hill. 1990. *Black feminist thought.* New York: Routledge.

Daniels, Jessie. 1997. *White Lies: Race, class, gender and sexuality.* New York: Routledge.

—. 2007. Race, civil rights and hate speech in the digital era. In *Race and Ethnicity*, edited by Anna Everett, MacArthur Foundation Series on Digital Media and Learning. Cambridge, MA: MIT Press, in press.

Everett, Anna. 2007. *Race and ethnicity.* MacArthur Foundation Series on Digital Media and Learning.

—. 2002. The revolution will be digitized: afro centricity and the digital public sphere. *Social Text* 20:125-146.

Feagin, Joe R. 2006. *Systemic racism: a theory of oppression.* New York: Routledge.

Festa, Paul. 2002. Controversial domains go to civil rights groups. *C|NET News.com* http://news.com.com/Controversial+domains+go+to+ civil+rights+groups/2100-1023_3-210803.html. Accessed June 28, 2005.

Gerstenfeld, Phyllis B., Diana R. Grant, and Chin-Up Chiang. 2003. Hate online: a content analysis of extremist internet sites. *Analyses of Social Issues and Public Policy* 3:29-44.

Glister, Paul. 1997. *Digital literacy.* New York: John Wiley & Sons.

Green, R. Michelle. 2006. Personality, race, age and the development of digital fluency. Paper presented at the American Educational Researchers Association Conference. San Francisco, CA.

—. 2005. *Predictors of Digital Fluency.* Ph.D. Dissertation. Northwestern University.

Hara, Noriko and Zilia Estrada. 2003. Hate and peace in a connected world: comparing Moveon and Stormfront. *First Monday* 8. 12 December. http://www.firstmonday.org/issues/issue8_12/hara/index.html (Accessed January 8, 2004.)

Jones, Steven. 1999. *Doing Internet research: critical issues and methods for examining the net.* Thousand Oaks, CA: Sage.

Kahn, Richard and Douglas Kellner. 2004. New media and Internet activism: from the 'battle of seattle' to blogging. *New Media and Society* 6:87-95.

Kaplan, Jeffery, Leonard Weinberg, and Ted Oleson. 2003. Dreams and realities in cyberspace: white aryan resistance and the World Church of the Creator. *Patterns of Prejudice* 37:139-155.

Kolko, Beth, Lisa Nakamura, and Gilbert B. Rodman (Eds.). 2000. *Race in cyberspace.* New York: Routledge.

Lamberg, Lynne. 2001. Hate-group web sites target children, teens. *Psychiatric News* 36:26.

Lee, Elissa and Laura Leets. 2002. Persuasive storytelling by hate groups online: examining its effects on adolescents. *American Behavioral Scientist* 45:927-957.

Lenhart, Amanda, Mary Madden, and Paul Hitlin. 2005. *Teens and technology: youth are leading the transition to a fully wired and mobile nation.* Washington, D.C.: Pew Internet & American Life Project.

Levin, Brian. 2002. Cyberhate: A Legal and Historical Analysis of Extremists' Use of Computer Networks In America. *American Behavioral Scientist* 45:958-988.

Markham, Annette. 1998. *Life online: researching real experience in virtual space.* Walnut Creek, CA: AltaMira Press.

Mills, Charles W. 1997. *The racial contract.* Ithaca, NY: Cornell University Press.

Mitchell, William J. 1992. *The reconfigured eye: visual truth in the post-photographic era.* Cambridge, MA.: MIT Press.

Mock, Karen. 2000. Hate on the Internet. In *Human Rights and the Internet,* edited by Steven Hicks, Edward Halpin, and Eric Hoskins, 141-152. New York: Palgrave Publishers.

Nakamura, Lisa. 2002. *Cybertypes: race, ethnicity, and identity on the Internet.* New York: Routledge.

Ray, Beverly and George E. Marsh. 2001. Recruitment by extremist groups on the Internet. *First Monday* 6, 2. http://www.firstmonday.org/issues/issue6_2/ray/index.html (Accessed January 24, 2003).

Resnick, Michael. 2002. Rethinking learning in the digital age. In *The global information technology report: readiness for the networked world*, edited by G. Kirkham. Oxford: Oxford University Press.

Rheingold, Howard. 2006. Keynote Address. NMC Online Conference.

Roberts, Donald F., Ulla G. Foehr, and Victoria Rideout. 2005. *Generation m: media in the lives of 8-18 year-olds*. A Kaiser Family Foundation Study. Menlo Park, CA: Kaiser Family Foundation.

Sherman, Chris and Gary Price. 2001. *The invisible web: uncovering information sources search engines can't see*. Toronto: CyberAge Books.

Smith, J.R. and Shih-Fu Chang. 1997. Visually searching the web for content. *IEEE Multimedia* 3:12-20.

Taft, Jessica. 2006. "I'm not a politics person": teenage girls, oppositional consciousness, and the meaning of politics. *Politics & Gender* 2:329-352.

Tateo, Luca. 2005. The Italian extreme right on-line network: An exploratory study using an integrated social network analysis and content analysis approach. *Journal of Computer-Mediated Communication* Vol. 10, Issue 2. http://jcmc.indiana.edu/vol10/issue2/tateo.html (Accessed June 28, 2006).

van Someren, Maarten W., Yvonne F. Barnard and Jacobijn A.C. Sandberg. 1994. *The think aloud method: a practical guide to modeling cognitive processes*. London: Academic Press.

Wellman, Barry. 2004. The three ages of Internet studies: ten, five and zero years ago. *New Media and Society* 6:123-129.

Whine, Michael. 2000. Cyberspace: a new medium for communication, command and control by extremists. *Studies in Conflict and Terrorism* 22:231-245.

The Hand Blocks the Screen: A Consideration of the Ways the Interface Is Raced[1]

Michele White

The arrow shaped cursor, or pointer, and hand are key aspects of Internet and computer interfaces. The arrow usually turns into a white pointing and clicking hand when "mousing" over web links and a white grasping hand when programs or images can be changed. Depending on the operating system and settings, white hands holding writing and drawing implements and other representations of hands also demarcate computer work. The hand moves when the spectator manipulates the mouse, relates the embodied individual who is in front of the screen to representations of bodies, locates the individual in the setting, and indicates that documents and links can be controlled, grasped, and touched. Representations of hands downplay the interface because the user seems to have slipped inside the screen, engages in a "hands on" way, and does such things as "hand code." The hand-pointer shapes our conceptions of the interface and Internet spectatorship but it has received very little critical attention.

Depictions of hands are the most common image of the Internet and computer user. Individuals become attached to these hands and empowered by them because they chronicle actions and options within the setting. However, these hands do not equally represent all individuals. They tend to be white, or white and gloved, and provide spectators with constant messages about what individuals who use the Internet and computer look like. Many literary hypertext authors and net artists also employ versions of white hands and render white users. There are also some digital art practices, including the work of Mendi + Keith Obadike, which suggest how racial inequalities are produced through technologies and propose alternative imaging strategies. Acknowledging the range of individuals, who employ the computer, and the ways they are addressed or ignored, requires a rethinking and redesigning of the interface as well as the ways we speak, think, and write about the Internet and computer.

In this article, I examine the hand-pointer, suggest connections to previous media representations and computer advertising, and consider how race is rendered through the interface. I suggest how white hand-pointers get conflated with the many representations of white users and indicate that the computer interface, Internet settings, and other aspects of the technologies and social processes can unfortunately work together to welcome white users and suggest that they are the expected participants. I employ such humanities methods as close visual and textual analysis, critical race studies, considerations of whiteness, and visual culture studies. My considerations of writings about the interface by such individuals as Anna Everett, Jay David Bolter and Diane Gromala, Steven Johnson, Lisa Nakamura, Jakob Nielsen, and Don Norman indicate that different perceptions of the interface persist. They also suggest that conceptions of the computer can have larger cultural effects. For instance, Everett analyzes her computer start-up message and its "Pri Master Disk, Pri. Slave Disk, Sec. Master, Sec. Slave," which indicates that some programmers choose to base Internet and computer culture on a "digitally configured 'master/slave' relationship."[2] Start-up messages and renderings of white hands, which appear each time the computer is employed, frame Internet and computer engagements and make Everett and others, at least temporarily, hold back from engaging.

Some spectators believe that "on the Internet nobody knows you're a dog"—as Peter Steiner's cartoon from *The New Yorker* suggests, personal information cannot be verified, and considering identity issues is unnecessary.[3] Steiner's assertion is supported by a variety of academic and popular texts that indicate the Internet is a place where race, and thus the challenging of racism, is irrelevant. In Virginia Shea's often-quoted "netiquette" guidelines, she indicates that Internet anonymity makes it impossible to judge spectators by their age, body size, class, and race.[4] Sadie Plant argues that the Internet provides "access to resources which were once restricted to those with the right face, accent, race, sex, none of which now need to be declared."[5] The Jargon File attributes hackers' "gender- and color-blindness" to their engagement with text-based communication.[6] Unfortunately, the belief that race and other aspects of identity do not matter in Internet settings allows whiteness to continue as the norm and discourages individuals from recognizing the racist representations that persist in Internet settings. Despite celebrations of Internet equitability, instances of intolerance continue and some individuals are barraged with requests for personal information. In reaction to these conflicts, Art McGee questions the claim that there is something the matter with being a dog or having a less normative identity position.[7] Imagining that individuals can be liberated from being a dog—or what is imagined as an imperfect position—perpetuates the value of traditional identities. As Nakamura suggests, the offer of new positions "to redress the burdens of physical 'handicaps' such as age, gender, and race produce cybertypes which look remarkably like racial and gender stereotypes."[8]

The interface and Internet are raced as white by the prevalence of white hand-pointers, references to hands on web sites, and the tendency to depict white users. White gloved interface hands, which are occasionally associated with Mickey Mouse, too easily reference vaudeville and blackface—where gloves helped produce evaluative distinctions between blacks and whites. The spectator's skin color will never correspond exactly to the white and pink colors of the interface hand but these depictions still reference whiteness, articulate what users look like, and enforce a racial inside and outside in Internet and computer settings. Dyer indicates that individuals "may not literally be white, yet a colour term, white, is the primary means by which" these individuals are identified.[9] As Elizabeth Grosz indicates, body and identity are rendered through varied forms of cultural writing and inscriptions, which include Internet and computer spectatorship, wearing clothing and makeup, and driving and identifying with a car—all of which shape and expand understandings of bodies.[10] For some, the white hand-pointer is easier to map onto their body-construct and identity. Nevertheless, Johnson suggests that the same conflation of body and technology is facilitated for everyone when the interface "shapes the interaction between user and computer" and the pointer becomes the "user's virtual doppelgänger."[11] While Johnson indicates that the individual's relationship to the interface is a key aspect of computer use, which allows the user to "enter that world and truly manipulate things inside out," this engagement also validates and materializes white people.[12] The hand-pointer acts as a kind of avatar or extension of the body and supports other renderings of the individual. It becomes "attached" to depictions of whites in varied Internet settings and to the bodies of white individuals.

In most Internet settings, the centered position of white men is deemed to require no explanation or additional details. Susan C. Herring, Inna Kouper, Lois Ann Scheidt, and Elijah L. Wright indicate how the privileging of political and knowledge blogs in print and broadcast reports and the blog roll links to A-list bloggers—many of whom happen to be white men—create a setting where a very narrow demographic is featured even though blogging is imagined to be democratizing.[13] Individuals who are white almost never write about their race in Internet settings, although this omission and their textual descriptions of such features as hair and eye color and photographic portrayals often position them. People self-presenting as African American, from the African diaspora, Asian, Latino/a, from an indigenous culture, and other people of color often state their ethnicity and race in descriptions. As Dyer suggests, the power of whiteness is secured by not seeming to be anything particular. While numerous advertisements and other representations of white users establish the presence and authority of white individuals, the weight of these devices and their production of larger narratives about race remain largely unaddressed. Since the interface is often understood as neutral, rather than producing particular narratives about computer technologies, such things as hand-pointers are identified as "triviata" or as part of a "mouse fetish" because of their connection to Mickey Mouse rather than as part of larger cultural narratives that privilege whiteness.[14]

Human computer interface researchers like Nielsen and Norman have argued that interfaces should transparently deliver information. Norman indicates that the computer should be "quiet, invisible, unobtrusive, but it is too visible."[15] However, displacing the ways the computer interface is understood and making design presumptions appear to be material realities elides biases and the conveyed ideas about bodies and identities. As Bolter and Gromala argue, if "we look *through* the interface, we cannot appreciate the ways in which the interface shapes our experience.[16] For instance, Cynthia L. Selfe and Richard J. Selfe, Jr. have described how "computer interfaces order the virtual world according to a certain set of historical and social values."[17] While analyzing Apple documentation in the mid-1990s, they noted that it included "a preponderance of white people and icons of middle and upper-class white culture and professional, office-oriented computer use."[18] They indicate that these devices, including the desktop metaphor, do not make the interface comprehensible to everyone but instead address a largely middle and upper middle-class white audience who is engaged with corporate and desk culture rather than individuals who are employed in such areas as service work, domestic care, and labor.

The hand-pointer derives from manuscripts', newspapers', and other print media's representations of pointing hands, which direct readers to articles and other materials of interest. When it is used in manuscript books and printed literature, the pointing hand is also known as a fist, hand director, indicator, indicator mark, index, manicule, and printer's fist.[19] It is also related to the yad, which means hand in Hebrew. The yad, or pointer, includes a sculpted hand and is used to read the Torah. Victorian scrap, which was subsequently collaged into books and albums, depicts white hands that are pointing and holding. Jewelry, from the Victorian to contemporary period, also depicts light-colored ivory and enameled hands directing individuals and holding objects. The hand-pointer is also related to yellow pages advertisements that began using an image of a hand in 1970.[20] The "let your fingers do the walking" yellow pages advertisements now also refer to Internet searching.

IBM's advertisements and promotional materials for the 1981 release of its personal computers, with a command line interface, connected white-gloved hands to the computer by using Charlie Chaplin's little tramp and his white-gloved hands to represent their product. By referencing Chaplin, IBM suggests that their computer can facilitate play and connects the individual and technology, including the process of upgrading components while wearing white gloves. Apple appropriated and represented Chaplin in their "Macintosh for the rest of us" campaign with his hand positioned like the hand-pointer in order to reference the graphical user interface and indicate the superiority of their products. Apple continued to use the figure of Chaplin in the "Think Different" campaign, which began in 1997. More recently, Apple has distinguished between the Mac and PC in a series of television ads. Nevertheless, they represent both platforms as white men.

Some operating systems provide the opportunity to change the pointer. For instance, Windows XP offers changeable suites of pointer images. While these consist of such schemes as "Hands 1," "Hands 2," and "Windows Black," all of them include a white "Link Select" hand. In some of these pointer schemes there are also images of white hands holding writing implements that indicate such things as "Precision Select," "Text Select," and "Vertical Resize." Some individuals argue that the hand-pointer needs to be white so that it is most easily seen but there are system schemes for Windows XP that deploy black arrows and OS X provides a black arrow that is outlined in white. All of these representations are easily seen, perhaps even more easily located than the white hand-pointer, since the edges of windows and background of documents tend to be white or a very pale color.

Formal properties and continued racial references connect hand-pointers to other representations of users. For instance, the pointer scheme images of hands holding writing implements are related to the desktop and web site icons that portray a similarly configured hand. Desktop icons with hands have been used to represent such programs as MacDraw, MacWrite, and MacPaint. They also have at least a formal relationship to M. C. Escher's *The Drawing Hands*. Pointer scheme and icon images of hands correlate the technology to the human and indicate that the computer facilitates cultural production, including artistry and authorship. They also depict what the individuals who use computer technologies look like.

Technology companies tend to feature white bodies in their advertising, particularly when these technologies are purported to enable the individual to produce rather than "just" listen and view.[21] These representations get conflated and attached to the representations of white hands through a number of processes. Adesso, which sells input devices, uses a series of three images to advertise its products and provide links to "Data Input Devices," "Handwriting Input Devices," and "Audio/Video Input Devices."[22] In each of these images, white familial relationships and unions are associated with input technologies and thus hand-pointers and, more generally, with the computer and Internet. In the middle image, a young white couple—dressed in a white gown and tux—are getting married. The image evokes leisure, family connections, futures, youth, and heritage and a relationship to newness since the image appears below the menu for "Whats New" [sic]. Despite this claim to newness, representing users as white and imagining that computer technologies facilitate connectivity, leisure time, and access to natural settings, with flowers and trees, are typical narratives. The kinds of families that might be facilitated by this connectivity and union are featured on either side of this wedding image. Adesso suggests that families are facilitated or even produced by shared time in front of the screen but they also code all of these arrangements as white.

Representations of white normative unions and families are also included, along with images of white technology workers, in advertisements for computer technologies and peripherals. Logitech has tended to associate its mice with businessmen who are depicted with technology, and a computer mouse, in the palm of their hand and on the move.[23] AOpen, which produces mice as well as other hardware, depicts white corporate and household users.[24] AOpen's opening Flash sequence includes images of a white hand balancing a tray that displays its varied technology products, a white child who holds his finger to his lips to indicate that the products are silent, a white woman and man at a conference table, and a white couple leaning against each other in a home. They snuggle against each other, the woman's head and arm resting in a submissive gesture on the man's shoulder. Their arrangement and relationship are literally framed by the computer. If the individual selects the "USA" link, which disturbingly conflates varied parts of the world by leading to a page that is titled "N. America & Latin" while English is still associated with "Global," "USA," "Netherlands," and "Asia Pacific," then the opening screen reveals two young very light skinned girls relaxing with a remote.[25] In these representations, not only is the technology user coded as white but the computer is also imagined to be part of the white reproductive nuclear family.

Computer advertising not only figures white individuals but also dresses them in beige and other light colors to produce a white and light world. Rays of light, which emanate from many of the people, convey the idea that technologies facilitate knowledge and epiphanies, and also seemingly whiteness and lightness. The AOpen image of a couple has haloes of white radiating from them. A Logitech image, which promises free shipping, has a man appearing from a pile of technology.[26] His body is thrust back into space, arms spread wide, and white light emanates from his torso and the laptop. These images evoke religious iconography—rays, halos, and glowing objects—and suggest that computer technologies have facilitated a new form of spiritual transcendence for the white individual. Such narratives are encouraged by early cyberpunk literature, including works by William Gibson and Neal Stephenson, which locate loas and other gods in the technology, although in these literary instances the race of users and their gods are more complicated.[27]

The advertising for mice and related input devices also tends to depict hand-pointers, icons of hands, and white users whose hands are positioned so that they match the formal arrangement of hand-pointers. Adesso's tablet advertisements also feature white models holding "pens." These representations connect the depicted hands that are featured in pointer schemes and desktop icons to the physical hand of the user. For instance, the representation of the hand in the listing for the Adesso CyberTabletM17 17" LCD Graphic Tablet Monitor is in almost the same position as the precision select pointer even though this means that the hand, and the body that is not depicted, would be at an uncomfortably peripheral position when working on the screen.[28] In this Adesso image, the hand appears to be both a rendering and flesh. It is cut off just beyond the edge of the tablet and thus suggests that the user is both inside and outside the screen and can touch and control representations.

The Wacom site is divided into "Americas," "Asia Pacific," "China," "Europe," and "Japan" but relates these different geographies to a white man and other images of white users. Wacom maps individuals onto continents, territories, and colonial histories because English is the only language listed for the "Americas" and is also offered as a selection for people in other parts of the world. Wacom depicts the man in a split screen image that divides his body from his hands. One of the images features most of the tablet and the man holding a pen so that it reproduces the pointer scheme representation of a hand holding a writing implement. By using this split screen device, Wacom represents the individual who is drawing, the embodied hand, and hand-pointer schemes as part of the computer interface. Like in the Adesso advertisements, Wacom conflates the white user with the hand-pointer and other screen-based icons of hands.

Advertisements for input devices depict white hands in control of the computer technologies—maneuvering the mouse and having a wide array of technologies literally and figuratively at hand. Bruce Tognazzini, who does human computer interface research, argues that such computer interfaces should instill "in their users a sense of control."[29] Websites for input devices and other hardware tend to not only address white individuals but also promise them a high level of control if they employ computer technologies. Software companies and Internet service sites provide similar narratives. For instance, Yahoo!'s web hosting icon depicts a white hand supporting a globe, suggests that the world will be under the white individual's control, and represents the accompanying business owner as a white individual who "can have it all!"[30] Print advertisements and television commercials represent a similar set of white interface hands in order to indicate the interactive power of readers and viewers. Programmers and active users tend to be depicted as white men. When software and hardware are magical or a breeze to use then they are often associated with women. An advertisement for Yahoo!'s instant messenger presents the users of the program, avatars, and audibles as white.[31] Another representation depicts two young women standing in the same space.[32] A woman whispers into the other's ear while talk emoticons hover overhead and indicate that their news is a secret. The emoticons, replete with white hands, correlate the interface hands to the actions and bodies of these individuals. The representation equates synchronous communication with gossiping, suggests that this is a woman's activity, and downplays women's employment of Internet and computer technologies.

Web sites offer much fewer representations of people of color working with computers than depictions of white individuals. Among the representations of people of color, Asian individuals are the most likely to be depicted engaged with technologies in authoritative ways. Men and women from the African diaspora, as well as white women, are more likely to be depicted in social and leisure settings and using such technologies as stereo systems. For instance, Logitech presents an image of a black woman posed with iPod speakers, which through her casual clothing (rather than the more corporate and upscale clothing in other advertisements) and the placement of the speakers on her shoulder reproduces cultural narratives about black people and boomboxes.[33] The woman listens to the music rather than producing content. Her rolled eyes and hugely open mouth evoke the performances that black people were called on to deliver in vaudeville shows, film, and television before the Civil Rights Movement and changing viewer demographics encouraged a rethinking of these cultural stereotypes. Melbourne S. Cummings indicates how black people in these staged settings "shuffled, rather than walked; they popped and rolled their eyes; they giggled" because "white audiences found these lovable, enjoyable, entertaining, and controllable."[34] In continuing such images, Logitech and other companies render black people as controllable rather than in control of technologies.

While many operating system designers, software developers, and producers of Internet settings downplay the produced aspects of the interface, Bolter and Gromala describe how Internet, computer, and net artists foreground the interface. They identify a corrective to "the assumption that the computer should disappear" in digital art.[35] For example, Garnet Hertz's desktop icons for Elmer's glue, fried rice, milk, red tape, and a myriad of other things playfully challenge the idea that Internet and computer settings materialize objects and desires.[36] However, his representations, as well as other icons, include the typical renderings of white hands. In Jeanie Dean's *A New Alphabet*, pointing hands are used to read the text and make it into a tactile process and a reconceived language but these hands also evoke the white user.[37] These representations of hands assist in making Internet settings seem like tactile and spatial worlds where real bodies exist. Despite Bolter and Gromala's descriptions of the critical work done by new media artists, digital artworks do not foreground all aspects of the interface. Instead, these artists' works are aligned with popular narratives about Internet materiality.

There are also artworks that do not facilitate the white interface. The menu for Mendi + Keith Obadike's *The Pink of Stealth* evokes the interface hand while suggesting that users have a range of racial positions and skin tones.[38] Their artwork considers how pink articulates class and race. It references films, Thomas Pink, hunting, the hexadecimal numbers for different pinks that are used in html programming, and the coding that underlies conceptions of race. *The Pink of Stealth* situates whiteness within an array of pinks rather than leaving the color white and white positions unquestioned. Despite such important critiques, the hand-pointer continues to morph into new configurations inside and outside of the screen. If the hand-pointer was once a remediation of manuscript, printing, and writing processes that was incorporated into the interface; it is now a symbol of technological and computer speed, power, and interactivity in print, television, and material settings.

The hand-pointer has been literalized and materialized as a pointer with a white or pink colored hand on it for giving lectures and demonstrations.[39] This hand-pointer has also been incorporated into varied children's toys such as Pretend & Play School Set and ABC Chalk Talk Electronic Learning Chalkboard. If, as FAO Schwartz and other sellers may suggest, the ABC Chalk Talk Electronic Learning Chalkboard "motivates learning through hands-on play" then the kinds of play and the color of the learning child's hand have been demarcated. Learning, pointing, computing, and knowing are associated with the white hand-pointer that is raised in the air and the white body that it references. Not surprisingly, the children depicted at play with these toys in web site advertisements are white.[40] Sellers may market the white hand-pointer as a device for "imaginative play and classroom participation" but aspects of these engagements have already been predetermined.[41] The hand-pointer provides another view of the computer hand-pointer outside of the screen and further connects white bodies to the white interface. Dyer indicates that whiteness needs to be made strange so that it can be identified and critiqued. In order for this to occur, we need to admit that white things are often correlated to white bodies, despite the tendency to detach whiteness from race, and foreground the connections between white hand-pointer schemes, icons of white hands, advertising images of white users, print and television references to white hands, physical white hand-pointers, and other representations of white individuals using computers.

Endnotes

[1] Some of the sites of analysis and theoretical arguments posed in this article also appear in my book and a related chapter. Michele White, *The Body and the Screen: Theories of Internet Spectatorship* (Cambridge, MA: MIT Press, 2006); and Michele White, "Black Is, Black Paint: Art Practices and the Erasure of Afrogeeks in Internet Settings," in *Afrogeeks: Beyond the Digital Divide*, ed. Anna Everett and Amber J. Wallace (Santa Barbara: The Center for Black Studies Research, 2007), 165-181. This article could not have been written without the support of the Institute for Advanced Study, the National Endowment for the Humanities, The Mellon Foundation, Newcomb College Institute, Tulane University, and Wellesley College. I owe particular thanks to Anna Everett, Ken Gonzales-Day, Kate Hayles, and Maggie Morse as well as all of the wonderful presenters at the first Afrogeeks conference.

[2] Anna Everett, "The Revolution Will Be Digitized: Afrocentricity and the Digital Public Sphere," *Social Text* 71 (Summer 2002): 125.

[3] Peter Steiner, *The New Yorker* 69, 20 (5 July 1993): 61. This statement and the cartoon are often reproduced on various web sites.

[4] Virginia Shea, "Core Rules of Netiquette," *Netiquette* (San Francisco: Albion Books, 1994), 40; and Virginia Shea, "Rule 5: Make Yourself Look Good Online," The Core Rules of Netiquette, 8 Apr. 2007 <http://www.albion.com/netiquette/rule5.html>.

[5] Sadie Plant, "Nets," in *Zeros + Ones: Digital Women and the New Technoculture* (New York: Doubleday, 1997), 46.

[6] The Jargon File, "Gender and Ethnicity," *The Jargon File 4. 4. 7*, 29 Dec. 2003, 8 Apr. 2007 <http://www.catb.org/~esr/jargon/html/demographics.html>.

[7] Art McGee, "Computers, Freedom and Privacy Conference," 1995.

[8] Lisa Nakamura, "Cybertypes and the Work of Race in the Age of Digital Reproduction," in *New Media/Old Media: A History and Theory Reader*, ed. Wendy Hui Kyong Chun and Thomas Keenan (New York: Routledge, 2006), 319.

[9] Richard Dyer, *White* (New York: Routledge, 1997), 42.

[10] Elizabeth Grosz, "Sexual Difference and the Problem of Essentialism," in *Space, Time, and Perversion* (New York and London: Routledge, 1995), 35.

[11] Steven Johnson, *Interface Culture: How New Technology Transforms the Way We Create & Communicate* (New York: Basic Books, 1997), 14 and 22.

[12] Ibid., 22.

[13] Susan C. Herring, Inna Kouper, Lois Ann Scheidt, and Elijah L. Wright. "Women and Children Last: The Discursive Construction of Weblogs," in *Into the Blogosphere: Rhetoric, Community, and Culture of Weblogs*, ed. Laura Gurak, Smiljana Antonijevic, Laurie Johnson, Clancy Ratliff, and Jessica Reyman, 2004, 6 Apr. 2007 <http://blog.lib.umn.edu/blogosphere/>.

[14] kaizen "Racing the Interface," *The Wired Campus*, 7 Oct. 2005, 7 Apr. 2007 <http://chronicle.com/wiredcampus/article/701/racing-the-interface>; and Lucy Bogan, "Racing the Interface," *The Wired Campus*, 7 Oct. 2005, 7 Apr. 2007 <http://chronicle.com/wiredcampus/article/701/racing-the-interface>.

[15] Donald A. Norman, *The Invisible Computer: Why Good Products Can Fail, the Personal Computer Is So Complex, and Information Appliances Are the Solution* (Cambridge, MA: MIT Press, 1999), viii.

[16] David Bolter and Diane Gromala, *Windows and Mirrors: Interaction Design, Digital Art, and the Myth of Transparency* (Cambridge, MA: MIT Press, 2003), 9.

[17] Cynthia L. Selfe and Richard J. Selfe, Jr., "The Politics of the Interface: Power and Its Exercise in Electronic Contact Zones," *College Composition and Communication* 45, 4 (Dec. 1994), 485.

[18] Ibid.

[19] On SHARP-L, which is sponsored by the Society for the History of Authorship, Reading, and Publishing, participants shared this list of terms and made connections between drawings of hands in manuscripts, the carrying over of this tendency in printed literature, and the computer hand-pointer. SHARP-L, "pointing hand," 30 June 2004, 4 June 2006 <https://listserv.indiana.edu/cgi-bin/wa-iub.exe?A1=ind0406&L=sharp-l>.

[20] Yell, "Yell," 20 Dec. 2004 <http://www.yellgroup.com/802569EA00621809/%2Aview%2A/C18FFB3551B6500A80256BF200340DA7>.

[21] Asians are more likely to be depicted using computer technologies than people from the African Diaspora, Latinos/as, or other people of color.

[22] Adesso, "Adesso --> Home," <http://adesso.com/>.

[23] Logitech, "Logitech – Leading web camera, wireless keyboard," 15 Aug. 2004 <http://www.logitech.com/index.cfm?countryid=19&languageid=1>; and Logitech, "Logitech Products > Mice and Trackballs," 7 Apr. 2007 <http://www.logitech.com/index.cfm/products/categories/US/EN,crid=2133>.

[24] AOPen, "AOPen Inc.," 6 Apr. 2007 <http://www.aopen.com/>.

[25] AOpen, "N. America & Latin," 6 Apr. 2007 <http://usa.aopen.com/>.

[26] Logitech, "Logitech – Leading web camera, wireless keyboard and mouse maker," 7 Apr. 2007 <http://www.logitech.com/index.cfm?countryid=19&languageid=1>.

[27] William Gibson, *Neuromancer* (New York: Ace Books, 1984); and Neal Stephenson, *Snow Crash* (New York: Bantam Books, 1992).

[28] Adesso, "Adesso --> Home," 3 Apr. 2007 <http://adesso.com/products.asp?categoryid=17>.

[29] Bruce Tognazzini, "First Principles of Interaction Design," AskTog, 2003, 5 Apr. 2007 <http://www.asktog.com/basics/firstPrinciples.html>.

[30] Yahoo!, "Web Hosting Services from Yahoo! Small Business," 23 Dec. 2004 <http://smallbusiness.yahoo.com/webhosting/>.

[31] Yahoo!, "Yahoo! Messenger," 20 Dec. 2004 <http://im.yahoo.com/>.

[32] Yahoo!, 7 July 2004 <http://yahoo.com>.

[33] Logitech, "Logitech Products > iPod/MP3 Accessories," 8 Apr. 2007 <http://www.logitech.com/index.cfm/products/categories/US/EN,crid=2407,categoryid=471>.

[34] Melbourne S. Cummings, "The Changing Image of the Black Family on Television," *Journal of Popular Culture* 22, 2 (Fall 1988): 75.

[35] Bolter and Gromala, 6.

[36] Garnet Hertz, "desktop_10jpg. 832x264 pixels," 4 June 2006 <http://www.conceptlab.com/desktop/img/desktop_10.jpg>; and Desktop Is, "DESKTOP IS *DESKTOPS*" 31 Aug. 2004 <http://www.easylife.org/desktop/desktops.html>.

[37] Jeanie Dean, *A New Alphabet*, 2003, 7 Apr. 2007 <http://pw.english.uwm.edu/~jdean/letters.html>.

[38] Mendi + Keith Obadike, *The Pink of Stealth*, 2003, 2 July 2006 <http://www.blacknetart.com/pink.html>.

[39] Nasco Reading Resources, "Overhead Hand Pointer – Large ~ Hand Pointer ~ Books & Resources ~ Nasco," 8 Apr. 2007 <http://www.enasco.com/ProductDetail.do?sku=TB18896L>.

[40] FAO Schwarz, "ABC Chalk Talk Electronic Learning Chalkboard at FAO Schwartz," 8 Apr. 2007 <http://www.fao.com/catalog/product.jsp?productId=5975>.

[41] Teaching Trends, "Teaching Trends – Resources for Teachers," 8 Apr. 2007 <http://www.teachingtrends.co.uk/acatalog/ler2655.html>.

Connecting the (Virtual) Dots

Simulations, emergence, augmented life, and visualization technologies animate cultural spaces, historical enterprises, games, and corpora as well as the military.

Timothy R. Tangherlini, Zoe Borovsky, and Todd Presner
"Thick Viewing: Integrated Visualization Environments for Humanities Research on Complex Corpora"

Helen Papagiannis
"Augmented Memories, Digital and Analog Realities"

John H. Johnston*
"Artificial Life: New Media Object as a New Space of Exploration"

Caren Kaplan
"'Everything is Connected': Aerial Perspectives, the 'Revolution in Military Affairs,' and Digital Culture"

* Available in digital format in the video archives of the conference at www.hastac.org.

Thick Viewing: Integrated Visualization Environments for Humanities Research on Complex Corpora

Timothy R. Tangherlini
Todd Presner
Zoe Borovsky

Visualizing digital assets—as well as the relationships between those assets—in humanities collections is a significant challenge. Properly addressed, the visualization of such assets (from hand-written manuscripts, to historical maps, to paintings, to buildings, to archaeological artifacts and beyond) and their interrelationships promises to allow scholars in both the humanities and the humanistic social sciences an opportunity to address long unanswered questions and to begin posing new questions to materials that have otherwise been well-described in traditional scholarship. Further developing various publication models derived from these well-structured systems holds the promise of bridging the divide between traditional print scholarship and more complex models of representation.

One of the main advantages of computing is the ability of machines to discover in a highly automated fashion patterns and relationships between assets in enormous corpora (many tens of thousands to millions of discrete assets), present those relationships in a visually meaningful manner, and allow end-users to make use of intuitive visual navigation to move within the collection. Beyond expanding the research horizon, the visual presentation of complex assets—or corpora of assets—also allows for wider access to otherwise inaccessible materials. Taken together—the ability of computers to discover, visualize and analyze patterned relationships, and the ability of the digital realm to increase access to such corpora with all of the implications of such access—presages a sea-change in scholarship in the humanities and humanistic social sciences. Our presentation here today focuses on two projects—Danish Folklore and Hypermedia Berlin—both of which present challenges related to computing and visualization because of the heterogeneous nature of the sizable corpora related to each project. Our closing remarks will highlight other UCLA projects, including the Digital Encyclopedia of Egyptology, that are equally engaged with these problems.

Danish Folklore

Danish Folklore is based on the enormous nineteenth century archival collections of Evald Tang Kristensen (1843-1929), the most prolific collector of folklore in Europe. The main component of his collections are 24,000 hand written field diary pages containing stories, songs, games, and descriptions of everyday life collected from nearly 7,000 named individuals. Over the course of his collecting career from 1867 until his death—a period that saw the move toward a democratic Denmark, the development of the railways, electricity and the telephone, as well as the motorcar, urbanization and the beginnings of the social welfare state—he also amassed a sizable collection of material items from rural life, corresponded voluminously with well-known intellectual figures including Grundtvig and Ibsen, and took hundreds of photographs. He encouraged others to collect folklore and descriptions of daily life and send them to him, thereby amassing thousands of pages of hand written manuscripts from ministers, school teachers and university students. Apart from collecting, he edited and published editions of his collected stories and songs, after making fair copy of excerpts of the field diaries. At the end of his career, he had published over forty-five volumes of folklore—some indexed, some not. Fortunately, he produced a four volume memoir of all his travels that included vignettes of most of the people he met, all based on his voluminous correspondence with his wife. Unfortunately, these memoirs were not indexed either. At the same time as he was undertaking this massive collecting enterprise—an enterprise in no small part conditioned by a burgeoning Romantic nationalism—the Danish state was deeply engaged in developing elaborate census data, taxation and probate records, and mapping the landscape, while institutions such as the Lutheran church and insurance companies were busy detailing the minutia of people's lives. All of these materials exist in various Danish archives, some in digital form and others not.

In short, the collection is an intriguing example of a remarkably complex humanities corpus—it not only includes the creative and scholarly output of a single individual (in this case Tang Kristensen), but it also includes the creative output of thousands of other individuals. Seen in this context, the collection is far more than simply a bunch of old stories. While Tang Kristensen's correspondence provides an intriguing window into intellectual and political debates of the time, his storyteller's narratives offer a fascinating lens onto changes in the social, political and economic organization of the countryside. The remarkably detailed historical maps produced at the time allow one to trace changes or discover phenomena in the physical environment that often lie at the root of a particular story. Ancillary materials such as census, insurance and church records contribute to the ethnographically thick description that suddenly begins to take shape when these records are placed in proper relation to one another. The biggest challenge of the project is making sense of this vast amount of data, and then structuring it so that computational techniques can help discover meaningful patterns. These patterns in turn can help us discern the complexities not only of traditional expressions and the politics of their collection, but also the nuances of everyday life in late nineteenth century rural Denmark.

Most folklore collections paint a remarkably one dimensional view of tradition focusing either on "typical" stories organized around themes and genre, or on the endeavors of a single collector. Scant if any attention is paid to the individual storytellers. The result of these standard presentations of folklore is that the complexity of the interrelationships between the collector, the storytellers, the social/political and physical environment, and their stories all disappear. The goal of the Danish Folklore project is to present a series of tools that allows one to visualize these interrelationships, easily navigate and access the underlying archival materials and, ultimately, understand the entire archive in an ethnographically "thick" manner. By connecting the materials to each other so that the original relationships between storyteller, story, collector, and environment are reestablished, the archive comes alive. Similarly, by connecting the collector to his collaborators, interlocutors, critics and family, and by connecting the storytellers to their social and physical environments through maps and state archives, the richness of these individuals' lives is much easier to comprehend. Tools that allow one to search across storytellers' repertoires reveal the interconnectedness of both of the storytelling tradition and the storytellers themselves. Finally, by incorporating visualization tools such as mapping, clustering and other word study tools, the archive opens up to new vistas for interpreting the archive.

The current project incorporates several main "views" onto this remarkably complex corpus. As more of the archival assets are digitized, such as Tang Kristensen's correspondence, more "views" will be added. The three main views onto the archive in the current project are the fieldtrip view, the informant view, and the story view. The fieldtrip view provides a map over one or more of the collecting trips taken by Tang Kristensen—the user can choose to focus on a single trip, or a series of trips, or all of the trips, by selecting map layers that describe the routes. Informants who he visited on these trips appear as icons along the mapped route. Clicking on an icon allows one to explore in greater detail an individual's life and her folklore repertoire. A description of the fieldtrip from Tang Kristensen's memoirs is accessible from this view and, ultimately, the correspondence that lies behind the memoir entry will also be accessible.

The informant view is the second main view onto the archive. This view maps not only the informant's biographical data into the local landscape using historical maps, but also maps all of the stories that he or she told into that landscape as well. From this view, one has access to a list of fieldtrips on which the informant was visited (bringing one back to the fieldtrip view), a biography of the informant and all of the archival material that relates to that informant (including census, church records, insurance rolls, enlistment rolls, and probate records). A link brings one to photographs of the informant and, in a very small number of cases, audio recordings of the informant originally made on wax cylinders. An index of stories, organized by fieldtrip and the order in which they were told leads to individual story views.

The story view includes the original hand written recording of the story, along with a Danish diplomatic transcription and English translation. The view also includes an image of the published version of the story, along with a Danish transcription and English translation. Finally, the view includes a scholarly annotation that in turn includes standard folkloric indices for the discovery of other story variants, as well as pointers to other variants in the collection. The underlying structure for incorporation and processing of the stories is still being developed. Ultimately, this view will incorporate various discovery tools including clustering, Sammon and Dendro-visualizations and the ability to perform lemmatized searches across texts. A final view into the archive will allow for discovery of informants or stories by place name.

Hypermedia Berlin

Hypermedia Berlin provides a far greater historical sweep than Danish folklore, and incorporates a greater diversity of assets, while it constrains its geographic scope to a particular city. The project (http://www.berlin.ucla.edu) is an interactive, web-based research platform and collaborative authoring environment for analyzing the cultural, architectural, and urban history of a city space. Through a multiplicity of richly detailed, fully annotated digital maps connected together by interlinking "hotspots" at hundreds of key regions, structures, and streets over Berlin's nearly 800 year history, the project brings the study of cultural and urban history together with the spatial analyses and modeling tools used by geographers. While all the historical maps are geo-referenced with latitude and longitude in order to perform spatial queries (such as mapping census data or performing longue durée comparisons), every map is preserved in its integrity as an epistemological record of the way in which Berlin was perceived, organized, and represented at a given time. The result is that the window or screen never becomes a portal of clarity, realism, or truth. Through the graphical user interface for Hypermedia Berlin, "the data dandy" (Manovich, 270) explores Berlin by zooming in and out of the maps, scrolling—in any order—through some 800 years of time, and clicking on various regions, neighborhoods, blocks, buildings, streets, and addresses. As the navigation is refined—both spatially and temporally—the database populates the search results with relevant media objects, which can, then, be viewed, selected, sorted, and recombined.

Analogous to the process of archaeological coring, then, data searches are bound by place (proximity) and time (duration), not simply keyword: A user might encircle a region extending, for example, fifteen city blocks south of Potsdamerplatz over the years 1920 to 1962. The data objects displayed in the results field are a function of the time-space coordinates determined by the user, what essentially amounts to a contingent narrative told from the database of possible elements. In this regard, Hypermedia Berlin responds to Manovich's challenge to consider the recursivity of database and narrative: "How can a narrative take into account the fact that its elements are organized in a database?

How can our new abilities to store vast amounts of data, to automatically classify, index, link, search, and instantly retrieve it, lead to new kinds of narratives?" (Manovich, 237). For the new media flaneur navigating through Berlin, a unique, hypermedia narrative is produced with each iteration, track, or pathway through the time-space database of the city.

But far from an information silo or "read-only" site, Hypermedia Berlin is constructed as a participatory platform with an elaborate tiered authorship component and a community annotation feature for generating content and data sets. Authenticated users—generally, those from the academic community—are able to add any sort of media object as well as select out material for courses and individual research projects. They also author and publish vetted multimedia articles using the resources of Hypermedia Berlin. General users of Hypermedia Berlin are able to add micro-annotations by geo-tagging points, lines, and polygons. The rationale is that micro-annotations contribute to the creation of a "people's history" of the city, leveraging the democratizing possibilities of the web to create, display, and distribute information. These annotations function as "folksonomies," which complement—but do not displace—the academically generated taxonomies or content, which is peer-reviewed and authenticated. Finally, Hypermedia Berlin leverages some of the new possibilities of the geo-spatial web by interfacing between the digital world and the physical environment. Because every object within Hypermedia Berlin is geo-referenced, a person equipped with a hand-held GPS device or even a GPS-enabled phone can both download and upload geo-specific historical information about their precise location. Through such location awareness technologies, a user standing in front of the Brandenburg Gate today will be able to automatically query Hypermedia Berlin for a 1962 picture of the Brandenburg Gate behind the Berlin Wall or view a map of the same location from 1811. The objective is to endow the Berlin of the present with its missing (or invisible) historical dimension. In this regard, the modern metropolis and new media begin to re-interface through a deep-linking dialectic: The metropolis changes new media, and new media changes the metropolis. As a kind of "augmented reality" (or, depending on what "side" one privileges, an "augmented virtuality"), the line separating media and the metropolis becomes blurred as Hypermedia Berlin is built on top, out of, inside of, and throughout the physical space of the city. In the present age of new media, the digital representational platform cannot be separated from the physical, geographic referent. This "new" new media thus moves significantly beyond first-generation web applications and content providers by combining a geo-temporal database with locative technologies, a participatory platform for community generated data, an interface between the digital and the built environment, and robust content created by extending and remixing publicly available interfaces (APIs).

Concluding Remarks: Visual Analytics: Connecting the Dots

Both Hypermedia Berlin and Danish Folklore are projects that attempt to create environments where large and disparate data sets can be presented in an integrated environment that allows users to perform a visual analysis of these materials. Both projects employ geo-spatial technologies--mapping their data onto representations of the material world. The advantages of adopting these tools as a way of exploiting the human eye's "broad bandwidth pathway into the mind" (Rhyme, 2006) for exploring and understanding large amounts of data simultaneously, seem remarkable. These projects support the promise that digital technology will provide new insights into materials that, by their sheer bulk and disparate nature, have not been presented in a way that promotes interdisciplinary scholarship and synthetic analysis.

The challenges facing humanities scholars who want to work in this fashion are strikingly similar to the agenda put forth by the National Visualization and Analytics Center that is funded by the Department of Homeland Security. Their materials cite the attack on the World Trade Center and Hurricane Katrina, as a "wake-up call" for scientists and technologists to formulate and carry out a research agenda for developing what they call "Visual Analytics": defined as "the ability to analyze large amounts of disparate data to make sense of complex situations and save lives." They are developing tools for visualization that will perform the following functions:

1. *facilitate understanding of massive and continually growing collections of data of multiple types;*
2. *provide frameworks for analyzing spatial and temporal data;*
3. *support the understanding of uncertain, incomplete and often misleading information;*
4. *provide user- and task-adaptable guided representations that enable full situation awareness while supporting development of detailed actions; and*
5. *support multiple levels of data and information abstraction, including the integration of different types of information into a single representation.*[1]

The reports from the National Visualization and Analytics Center contain a description of challenges similar to the ones these humanities projects face: short or long textual documents in many languages; numeric sensor data; structured data from relational databases; and audio, video and image data.

When we approach our deans to ask for funding for digital humanities, we are sometimes told that the funding must go to disciplines that can promise to save lives. Will taxpayers sleep safer knowing that stories told by Danish peasants are being scrutinized by folklorists? Are our borders more secure for having understood the political, social and cultural consequences of the Berlin Wall?

We may hesitate, for political reasons, to connect the dots between digital humanities and a science that presents itself as a defense against terrorism. Drawing such a comparison, given the sophistication of the tools the NVAC is developing and what we are using, is quite a stretch. However, I encourage humanists to study the reports, representations and tools being developed by this group and others. It is clear that decision-makers of the information age will rely on these tools and representations to inform their decisions. As humanists we can test similar tools and methods on materials we are familiar with, comparing the results with the outcome of more traditional analysis. In this way we can gain an understanding of how the tools may shape the outcome, leading to a critical assessment of the tools and the representations they produce. We can train ourselves and our students to engage with these tools and become familiar enough with the medium that we can make significant contributions that will shape the discourse of visual analytics in ways that will allow us all to sleep safer.

Endnotes

[1] A visual analytics agenda Thomas, J.J. Cook, K.A. Pacific Northwest Nat. Lab., Richland, WA, USA.

This paper appears in: Computer Graphics and Applications, IEEE
Publication Date: Jan.-Feb. 2006
Volume: 26, Issue: 1
On page(s): 10- 13
ISSN: 0272-1716
INSPEC Accession Number: 8735471
Digital Object Identifier: 10.1109/MCG.2006.5
Posted online: 2006-01-10 20:09:53.0

Augmented Memories, Digital and Analog Realities

Helen Papagiannis[1]

My current work in Augmented Reality (AR)[2] explores integrating AR markers with lenticular-based lenses. My intent is to create tactile objects that may store and display multiple moving AR images, combining both analog and digital modes of memory. I have always been mesmerized by the technology embedded in lenticulars and their ability to contain and reveal multiple images with a slight shift of hand. I have recently created an AR marker contained within a lenticular lens that presents two separate marker patterns. Each of these patterns reveals a different moving AR image when the lenticular object is slightly tilted. The end result is a layered form of a futuristic moving image, one which comes to exist via an analog mode of animation.

I have been experimenting with various applications for lenticular-based AR, one of which explores the ability to display memories over time from past to present, combining both archival footage with contemporary moving images. This technique may be used to show growth over time, or various stages of one's life memories. A recent lenticular AR prototype I have created first displays a black and white film clip of two children playing and shyly kissing each other on the cheek; the second marker reveals a video clip of the two children now grown-up playfully behaving in the same manner as they once did, viewed in the previous moving image. The lenticular based AR markers may be used to display a before and after of sorts. The viewer can 'flip' between the two moving images in the same hand-held object, mid-clip, reverting between the two, crossing over time with a slight hand gesture. Another prototype demonstrates the ability to change the direction of the moving image, between forward and reverse, when the hand-held lenticular object is slightly shifted.

I am particularly interested in the dual memory of the physical object and virtual imagery in lenticular-based AR. Although the augmented image is stored digitally within the software, activated upon recognition of the AR marker by the computer, the lenticular lens also contains an analog based memory system to store and reveal the two different markers with a physical tilting gesture. Each technology, AR and lenticular, presents an architecture which serves as a memory container with the final image only coming into full-view upon activation by the user. The completed images otherwise remain hidden from the viewer; the AR digital image appearing just as a square marker to the human eye without the software, and the lenticular analog image only a sole static still, unanimated. Although the AR image output is reliant on the software to translate and produce, the AR markers are initiated by the physical maneuvering of the lenticular lens by the viewer. This same gesturing is used to navigate between the final imagery, back and forth between the AR moving images. Both analog and digital methods must work together and coexist to bear lenticular-based AR. The direction of my current and future work looks to combine these two methods, utilizing both to create a final output where the digital and analog coalesce.

My work in AR began with a series of memory albums and paper-based objects which presented digital video footage from my travels. My interest in creating these works was due in part to a desire to capture 'live' moments from my sojourns that were beyond still photographs, which would aid to temporarily transport me back to these foreign locales to relive those instants. These moving images assisted to evoke and recollect my memories by being able to rearticulate a past vision of a particular location: once again seeing how the waves crashed, how the wind blew, how my body moved in a space which I no longer have physical access to. I found that unlike the digital photographs that I took and would eventually print and place in an album, these moving images (MPEG format) most often remained archived on disc or on my computer never to be experienced again. I desired to create a tactile object where I could hold and view these 'live' moments again, alongside my still photographs, offering an opportunity to move through the still images, extending into and beyond their virtual viewing space.

I created a series of small hand-held AR objects including a palm-sized memory album, a set of paper slides cased in a petite box, and a travelogue, which alongside video-clips, included actual objects from my journeys in addition to hand-written stories accompanying each clip. None of the moving images I chose to include featured people; they were all pans of landscapes of the sites I visited. I viewed this as an opportunity to document the physical places I visited, as a form of souvenir that would allow me to visually revisit (virtually) and enter that space again via a moving image that captured my field of vision in a horizontal pan. Without other people in the footage, this aided to create an intimate, uninterrupted space, as though that particular moment was for me, undisturbed by anyone else, a private memory, between that place and I. My works further exhibit a level of intimacy in their miniature scale; most of my projects fit in the palm of the viewer's hand.

*QuickTime videos and images of the work are available at: www.aliceglass.com

Endnotes

[1] (QuickTime videos of my work discussed below may be viewed at: http://www.aliceglass.com/research.html).

[2] Augmented Reality (AR) is the convergence of the real and the virtual, often consisting of the overlaying of computer graphics onto a physical environment, which is interactive in real-time. The form of AR technology I am presently working with is based upon a series of black and white square markers. A web camera is utilized to capture images of the real world, which is then sent to a computer. Software on the computer searches through the live video stream for the various square markers. Once the software has recognized an AR marker, the marker is replaced with the corresponding video file to create the final output, which is overlaid onto reality.

"Everything is Connected": Aerial Perspectives, the "Revolution in Military Affairs," and Digital Culture

Caren Kaplan

Since the middle of the 20th century, the environment above the earth's atmosphere has been, increasingly, militarized. Although space has been perceived as devoid of national or even worldly interests, throughout the Cold War and its aftermath, U.S. proponents of the military purposes of space argued convincingly for the development and expansion of communications and intelligence satellite programs (Stares 1985). With the breakup of the Soviet Union at the end of the century, the space "race" slowed, leaving the U.S. the predominant nationalized military presence in space (although China and the European Union have ambitious satellite programs). Weaponization and militarization of space are not exactly the same thing and in my comments today I will not be discussing the debate on weapons programs that rely upon space platforms or the various anti-satellite programs. I will be focusing my comments on the so-called "Revolution in Military Affairs" (RMA) and the ways in which contemporary, popular, post 9-11 representations of satellite surveillance-linked warfare are produced, distributed, and consumed. The RMA proposes a "network centric" military, drawing on information technologies and concepts such as the "system of systems." While digital culture is fairly new, there have been a number of so-called "revolutions" in military technologies and strategies over the last several hundred years (Hirst 2001). In the longer project of which this paper is a part, I am exploring the visual culture of militarization in the 20th and early 21st centuries, in particular "views from above;" that is, the emergence of photography, cinema, and war as linked technologies constituting new ways of seeing. The advent of aviation followed by the exploration of space and satellite programs has brought new powers of observation to the mechanical capturing or recording and reproduction of images (Virilio 1989). The rise of intelligence or reconnaissance photography—first from airplanes and then from satellite platforms—has its own detailed history in the study of air and space power that I can only gesture towards today.

The aspect of this project that I am exploring here concerns the popular culture of reconnaissance warfare as a prosthetic practice of air and space power, especially in the post-9-11 context of a crisis in military strategy. In some related, recently published work, I have engaged the argument that the military, business, and the entertainment industries are closely linked, even networked, in terms of their politics and culture as well as their economics (Kaplan 2006). As a cultural critic, I am interested in the culture end of this problematic—that is, how do you produce critique when various entities that appear to be distinct are, in fact, quite closely related? In this regard, for today, I want to focus on three examples of discrete but related representational practices that produce globalized subjects of US nationalism, all drawing on the discourses of networked societies that are heralded in the so-called "Revolution in Military Affairs" (RMA).

Just a few short years ago, the RMA, as touted by its most visible proponent, Donald Rumsfeld, symbolized the "network centric" evolution of technology at the turn of the new century. The war in Iraq was to be the best example of this linked network approach where "everything is connected," air power and space power would blast the nation's enemies to smithereens without jeopardizing ground forces unduly, and information technologies would solve many of the problems that have plagued militaries for centuries—accurate, real-time communications, precision delivery of armaments to targets, and moving people and supplies around in a timely manner. In short, as Mike Davis has put it, the Pentagon aspired to work like Wal-Mart (Davis 2003). We all know what happened to the RMA (and to Donald Rumsfeld!)—the RMA survives the insurgency in Iraq and Afghanistan but in a completely different context—as the vertical geopolitics of the current conflict reflects, this war requires ground troops and urban house to house fighting rather than air power in a de-populated desert. Nevertheless, the representations of RMA-type networked systems of vision and weaponry persist and even flourish in popular culture.

My first example is 30-second tv spot produced by aviation giant, Boeing Corporation, in 2003 as part of the Boeing "branding" effort that accompanied restructuring, executive reshuffling, and the relocation of corporate headquarters from Seattle to Chicago. The mega advertising agency, Foote, Cone, and Belding devised the slogan "Forever New Frontiers" to promote the company at the turn of the new century. This ad in the series is entitled "Bigger Picture." I have transcribed the text for the ad; it runs as follows:

Today a soldier sees a snapshot of the terrain
A pilot an image of the air space
The commander a view of the mission
But they're all part of a bigger picture
That's why Boeing is helping create a remarkable network

To gather and analyze data from every source
Then deliver the right information instantly
So the bigger picture is a safer world.

In this very artfully executed advertisement the world that is evoked is clearly post 9-11 and viewers are being hailed as subjects of the United States. Other ads in the series, some from before 2001, are more overtly "multi-culti" and cosmopolitan, featuring clips from diverse nations and cultures that are all presumably brought together by the kind of world that Boeing aircraft and time-space compression make possible. In this ad, the scope of the matter is very much the war that has been launched in 2003, the same year that the ad appeared. The "bigger picture" of the ad's title can be seen as an animation of points of view based on scale. As the commercial begins we hear the wind blow in a desolate, desert-like landscape, and "a soldier sees a snapshot of the terrain." This leads to the next scale, that of the pilot, who sees "space" and the commander, who sees the whole "mission." We move from the soldier's vantage point on the ground, to the aerial perspective of the fighter pilot, to the panoptic mastery of the commander, all in the service of the "bigger picture"—a more secure world (read "nation") via the linked media of surveillance, interpretation, communications, and targeting—that returns us back to the point of view of the individual soldier. The ad ends with a close-range view of the soldier gazing directly out at us, the viewers.

Figure 1. Boeing Corp. 2003. *The Bigger Picture*

My attention was caught by this ad in particular for two reasons: the representation of aerial perspectives as well as the discourses of a seamless network that dovetails so closely with the aims of the reformers in the military who have been agitating for the dominance of high-tech platforms. In the "Bigger Picture" tv spot, like so many of its ads, Boeing positions itself as a provider of a dream system—instantly, and seamlessly, multiple points of view and innumerable sources of information are integrated to slip back and forth between joint command centers and individuals and units in the air and on the ground. Everything is connected. It's a perfect illustration of the first two key points that Paul Hirst identified as the foundational claims of the RMA:

1. new information technologies will all but eliminate the so-called "fog of war"
2. RMA new technologies will lead to two opposed but complementary tendencies: the elimination of hierarchy, and the ability of senior commanders to enjoy total control of the whole battle space…

Many commentators have pointed out that such visions of technological derring do are highly overstated. A quick scan of the military enlistee blogs as well as the debates in such venues as *Air and Space Power Journal* reveal that not much is ideal. Human error combined with weather conditions and the state of the technologies themselves lead to many complications and challenges. Or, as Hirst put it, "information-centric systems create their own forms of fog" (Hirst 2005). But the fantasy that you can push a button in one place, say Washington, D.C., on any given moment, and that the satellite will be in just the right place and not a cloud will be in the sky, and that you can effect the old bombadier's dream of dropping the "pickle" right down into the pickle barrel of a target from 10,000 feet is dearly held by both conservatives and their critics. Here is my second example for our discussion, a clip from the film *Syriana*—the George Clooney/Matt Damon vehicle, written and directed by Stephen Gaghan, released by Warner Bros. Pictures in 2005 (Gaghan 2005).

Just a few words for those of you who may not have seen it—the film is based on the memoir of a former CIA operative in the Middle East and the main character travels from a solid foundation, a binary world view, to an understanding of the impossibility of fixing good and evil to national or even individual identities. As in all good spy narratives since the Cold War, the hero is torn, his world becomes ambiguous to the point of horror, and his existential dilemma becomes expanded to those of all good liberals viewing the film. In the scene I am going to show you from the very end of the movie, Clooney's character is racing to try to save the life of the Arab prince who symbolizes any last hope for democracy and humanist values in the region. For convoluted plot reasons, involving a very evil transnational oil company and corrupt politicians and intelligence officers back in the US, the decision has been made to assassinate the prince by long-range missile.

Figure 2. Publicity Poster. *Syriana*. Warner Bros. 2005.

Syriana's advertising campaign is the slogan "everything is connected." In the clip we just saw, that phrase resonates on several levels. Certainly, in the paranoid world of the political thriller, everything IS usually connected. For the pleasure of plot junkies everywhere, often, as the hero's life flits before him at the climax of the story—he suddenly puts it all together—it all falls into terrifying place. In this instance, we see the <u>prince</u> "get it" just before he "get's" it—just before he and his family are blasted to smithereens by a precision guided missile.

There is a lot to say about this clip and I can only point to a couple of things for now. One structural element I notice throughout the film that is extremely evident in the scenes we have just watched is the privileging of the direct gaze from the naked eye, apparently unassisted by technology, as the hope for human communications and "democracy." Thus, we see Clooney's character, spying the caravan of cars from a distance by eyesight. In this film, as in many others, the point of view from the ground, at the scale of the human gaze or touch is presented as the pov of democracy and humanism. This point is choreographed most powerfully in the editing of the look of recognition between the prince and Clooney's characters as the long-range missile closes in on them. At arm's length, almost touching, recognition and connection appear to be possible yet doomed, by the dehumanized gaze of surveillance from on high.

I am also struck by the representation of the seamless network of information and weaponry in this film and these representations bear a striking similarity to the Boeing ad. In the current zeitgeist, we are supersaturated with such images and scenes. The image I am showing you here is a composite put together by *The New York Times* showing aerial reconnaissance images of the precision missile strike that assassinated Abu Musab al-Zarqawi last June in Iraq. I don't have time to show you the clip of that strike but it is readily available on the internet, often set to stirring patriotic music or thundering rock and roll by pro-war YouTube enthusiasts.

Figure 3. *Syriana*. Warner Bros. 2005

These kinds of highly publicized incidents reinforce the belief that satellite surveillance is a constant and that a missile can be correctly navigated in an instant and the target will be precisely and completely destroyed. In *Syriana* the shot of the guy's thumb just above the joystick "trigger" as the "commander" in what seems to be the CIA says "take the target out" refers directly to the fantasy of a total system: you "see" as far and as high as need be with a satellite, you aim remotely, and you "shoot" your weapon with extreme speed and accuracy—the implication is that you can see anything, anywhere, and act or not act as you choose vis-à-vis anywhere in the world at any time, all from the comforts of your Aeron chair at Langeley. It's interesting to see this kind of "advertisement" for the Revolution in Military Affairs from a film that has been vilified as anti-Bush and anti-American. But this kind of representation of warfare is visually compelling and extremely popular in the post-9-11 crisis in security. Thus, my last example is a clip that has been making the rounds of popular culture video downloads. I found it on YouTube but you can find it just about anywhere online, usually along with a very interesting set of comments. The YouTube title of this clip is "F-16 strike on insurgents in Fallujah Iraq."

Figure 4. "F-16 strike on insurgents in Fallujah Iraq." n.d.

The final line we just heard, "oh, dude," is another title for this clip as it makes its way around the world wide web (or "aw, dude" is another way it is heard). The narrative of this clip echoes the pseudo-realism of the Hollywood product, *Syriana*. In that film, the guy in charge says "take the target out." We hear someone say "Roger." And then, "the target is destroyed." In the clip of the F-16 strike, we hear this:

> I got numerous individuals on the road, do you want me to take those out?
> Take 'em out.
> Ten seconds.
> Roger.
> Impact.
> Oh, dude.

The emotion implied by the comment "oh, dude" is echoed in remarks left on at least 3 posts on YouTube that display one version or another of this clip. While many comments run along the lines of "great video dude!" or "don't fuck with USA or England!" there are quite a few that ask "how do we know those people were insurgents?" Quite a lot of invective is heaped on those quiet queries but the question is a good one even as the debate rages over whether this clip really comes from the first Gulf War rather than the 2nd or whether it is a bunch of sheep in the road or a group of village elders, etc., etc. I am not interested in this instance in whether the clip is authentic or not because I think it is clearly connected to a repertoire of representations that propose the view from the sky as completely effective for weaponry. That is, once you've been "seen" from above, you're toast.

But as I've argued recently on other occasions, it's a little too easy to pose the orbital view as more lethal and less humane than the grounded or located scale of the naked gaze (Kaplan 2006). As a scan of numerous videos made by U.S. military personnel involved in the current warfare that get posted on sites like YouTube or other Hollywood products like *Jarhead* easily shows, the Marine sniper is a highly idealized trope in the Gulf wars (rather than the pilot) and the scale of conflict that is most often depicted, with accompanying thundering rock scores, is almost always hand-to-hand combat, or a view from a vehicle on the ground. The numerous shots of Iraqi or insurgent bodies with heads blown off at close range are as popular as the many homemade versions of videos that use the aerial views of the assassination of Zarqawi. They're all pretty violent and they all celebrate or propose the triumph of U.S. military prowess on all kinds of levels: hardware, software, muscle, eyesight, mobile phone snapshots… you name it. "Oh, dude" with its aerial perspective, its powerful dismissal of ambiguity—are they a herd of sheep, a bunch of civilians, a unit of terrorists…? Who cares, take 'em out! We're at war! is as popular as a video of guys in heavy armor breathing heavily running up a flight of stairs using night vision goggles to bust in the door and hold guns to the heads of the sleepy occupants. All of these images are circulating in popular culture. My point is that they can be differentiated according to specific genealogies and particular forms of discipline, governmentality, and other kinds of power.

Satellite surveillance still offers a special kind of power and meaning and we should be alert to its representational histories and effects. Those who seek investors and clients for aerospace technologies will draw on this fund of images and discourses for commercial gain. The military is, perhaps, much more complicated—full of debate and passionate differences of opinion, the US armed forces are struggling to unify their doctrines, strategies, and practices in the face of a very long war. The government is just as complex and shape shifting—just when you think it is never going to change, something happens, things come together or fall apart, and we find new configurations of power and profit at work. It is, maybe, the entertainment industry with its allied marketing and media practices that demonstrates the links in the "system of systems." Connecting these powerful entities to examine their representational practices is not to argue that they are hegemonically in lock-step as if in some kind of conspiracy. Rather, looking at the popular culture of militarization brings into sharp relief the "bigger picture" of advanced capitalism, the worlds it has made and unmade, and the views that produce the subjects of US empire.

Works Cited

Davis, Mike. 2003. War-Mart: "Revolution" in Warfare Slouches Toward Baghdad. *San Francisco Chronicle*, Sunday, March 9, D 1.

Gaghan, Stephen. 2005. *Syriana*. USA: Warner Bros. Pictures.

Hirst, Paul. 2001. *War and Power in the 21st Century*. Cambridge, UK: Polity Press.

———. 2005. *Space and Power: Politics, War and Architecture*. Cambridge, UK: Polity Press.

Kaplan, Caren. 2006. Mobility and War: The Cosmic View of U.S. Air Power. *Environment and Planning A* 38 (2):395-407.

———. 2006. Precision Targets: GPS and the Militarization of U.S. Consumer Identity. *American Quarterly* 58 (3):693-714.

Stares, Paul B. 1985. *The Militarization of Space: U.S. Policy, 1945-1984*. Ithaca: Cornell University Press.

Virilio, Paul. 1989. *War and Cinema: The Logistics of Perception*. Translated by P. Camiller. London: Verso.

Innerspace and Interface

Affect and representation are crucial to digital history, music, and dance.

Jennifer E. Boyle
"The Hollins Community Project: Interfacing Affect"

Tim Wang and Ulrich Rauch
"An Arts Metaverse: Reconstructing the Past (A Short Review)"

John Toenjes and David Marchant
"Finding Humanity within the Machine: Large Motor Movement Computer
Interfacing as an Artistic Mindbody Integrative Practice"

Interfacing Affect: The Hollins Community Project

Jennifer E. Boyle

Background to the Project

Hollins University, a private, women's liberal arts university located in southwest Virginia, is an embodiment of Foucault's *heterotopia*: the juxtaposition of seemingly incompatible notions of "real" and "imagined" spaces enfolded into a specific context of "place."[1] Of course, while Foucault gave the term new critical life, he did not invent it. Indeed, if one were to invoke the phrase at a medical conference on brain or body imaging science it would designate the displacement of embodied phenomena. In many ways, the new media project I describe in this paper brings together aspects of the two prevailing definitions of a heterotopia. *Interfacing Affect: The Hollins Community Project* is an experiment in new media that explores the ethics of new media interfaces through the emplacement of history, narrative, and embodied affect.

The planning for this project was initiated early in 2006 as part of a collaborative National Science Foundation proposal that linked the resources and faculty of the Center for Human-Computer Interaction at Virginia Polytechnic and State University (Virginia Tech) with a much smaller women's university, known for its unique arts and humanities programs. The collaboration between Hollins University and Virginia Tech brings together two institutions that are not only very different in size and mission, but with very different relationships to the history of southwest Virginia. Hollins in particular is an institution that has formed its notions of "community" out of an institutional identity grounded in "tradition" and local, self-generative history (many of the narratives describing the unique character of the University draw connections between the cultural and physical "environment" of Hollins and its "creative" aspirations and successes). In addition to being a well-ranked liberal arts university, Hollins also holds the distinction of having been in existence through both the prebellum and antebellum South (founded as a seminary in 1842, Hollins became a women's college in 1851). The physical topography of Hollins is itself a ghostly heterotopia – a rural landscape that juxtaposes state-of-the-art facilities and "literary landmarks" with ancient foot trails and a "grand old house on the hill."[2] On one end of campus is a small trail that leads through a small wooded area. The trail, located at the upper eastern edge of campus, is one of the few remaining geographical links to the Hollins Community, a term that refers to the community established by African American slaves who were brought to Hollins in the early eighteenth century and who remained in the area as "servants" to the university after Emancipation. The trail is a remaining trace of two sides of Hollins' institutional identity: a creative, progressive women's liberal arts institution with a strong community ethos, and the first chartered women's college in southwestern Virginia possessing both prebellum and antebellum histories. The trail itself is a fascinating confluence of the real and the virtual. Along the trail one finds discarded cookware and turn-of-the-century glassware; old small foundations for dwellings; the ruins of wood sculptures from the contemporary era. The trail and the environment surrounding it have had many afterlives: foot trail connecting members of the Hollins Community to the campus; a university dump site earlier in the 20th century; and a sculpture garden for Hollins art students.

A key influence for this project, Ethel Morgan Smith, author of *From Whence Cometh My Help*, has produced a history of the Hollins Community that mixes extensive archival and field research with personal reflections and creative non-fiction. The rich context she provides for the trail's history plays a fundamental role in both phases of this installation – from oral histories from members of the Hollins Community who describe leaving stones along the trails so their children could find them at work, to descriptions and images of the psychical, social, and geographical complexities of traversing between two worlds at once spatially and culturally divided and intimately intertwined.

The Hollins Community trail elicits productive confusion over the interrupted logics and narratives of institutional memory. The site is a montage of physical ruins and the competing and conflicted histories that seep through the texts, images, dirt, and artifacts associated with this small stretch of land. This project was conceived as a way of further intensifying the conjunction of these registers through the use of new media. New media has become a contested term for the progressive materiality of new digital technologies.[3] Our hope is that this project will serve as an example of how the thoughtful employment of new media interfaces can intensify spatial and temporal experience in such a way that *the incipient* moment of forming history in narrative can be experienced as embodied phenomena.

What are the possibilities for encountering alterity through new media technologies and interfaces?

"The ruin does not supervene like an accident upon a monument that was intact only yesterday. In the beginning there is ruin. Ruin is that which happens to the image from the moment of the first gaze." --- Jacques Derrida [4]

Derrida's musing on the ruin offers up an important critique of originary experience. Originary experience is constituted as a faith in a mythical pre-moment where history was once coherent and intact. In fact, the moment we enact a gaze with the hope of creating a personal or historical narrative out of an artifact, image, or space we are already inscribed in a process of decay. Other possible sites of meaning must be interned as a history emerges. Memory becomes de-composed to the extent that narrative or artistic composition must be adept at animating memory through framing, forgetting, and selecting what gestures, artifacts, and meanings will remain as representations of a given event.

The power of Ethel Morgan Smith's narrative experimentations in telling the history of the Hollins Community adheres in her insistence on retaining the unstable power of the ruin. While she gives "voice" to the African American community associated with Hollins, she imagines those voices as already embedded in the complex layering of her present narration and embodied experience at Hollins, the subaltern oral histories of former Community members, and the nuances of physical space, institutional memory, and narrative emplacement. While Smith's narratives and fragments construct a montage of ignored and invisible faces and histories, they do so with the intention of imaging them as agents of their own historical experiences.

The challenge for this project is to maintain and perhaps even productively intensify the ruptures and fragments to the narrative logic surrounding the Hollins Community Trail and the communities and histories associated with it. A principal claim informing these efforts, then, is that there is potential for new media interfaces to enact an affective (ethical) dimension to the instability of historical narrative.

Theoretical and Practical Components of the Installation

Several students from Hollins University will be traveling between Hollins and Virginia Tech over the summer and fall of 2007 to work on constructing the interface technologies for the project. Students and faculty will the return to the site at Hollins to experiment with specially programmed hand-held devices.

Students will be in residence at the Center for Human-Computer Interaction at Virginia Tech to learn how to work on programs for the handheld devices they will carry with them at the site. The objective in the design phase of the interface is to create an opportunity for students to think about how technologically coded environments open up or potentially limit interaction with the found objects at the site. The emphasis here is on embodied, performative inter-facing with these artifacts, rather than an augmentation or complete interruption of representational meaning.

The second phase of the installation asks participants to go to the site in small groups, equipped with handheld augmented reality technology. Drawing inspiration from Mark Dion's *Tate Thames Dig* [5], this segment of the project requires participants actively to locate, interact with, and "taxonomize" remnants, ruins, trash, fragments, and found objects along the Hollins trail. A central technological element of the interface involves creating traces (temporal spatial, and narrative) that record embodied movement and interaction with the objects. The user's physical movement in relation to their interaction with a given object is recorded; however, the user can also make their own choices about when to record a particular gesture or movement. Users also generate narrative and image content that become affixed to the time and spatial coordinates recorded. The patterns of activity that emerge are fragmentary and represent multiple dimensions of experience with the space: time/date traces that can be manually triggered; rough geometrical outlines of activity around a particular artifact; text and images produced as an interaction unfolds. This aspect of the installation is designed to evoke a sense of incipient action in relationship to images and artifacts that would seem over-determined if fully contextualized. The hope is to enact various encounters with a version of what Mark Hansen has described as the "digital-facial-image" (DFI) but through artifacts, gestures, and objects that in turn pose their own challenges and questions. The "digital-facial-image" is a circuit that moves perception of the image into a mode of affective interaction with and through the image. Such interfaces, "rather than channeling the body's contribution through a… narrow frame of software options… open the interface to the richness of bodily processing of the image."[6] The DFI circuit elicits recognition at a very basic level that the body's experiences are not a "closed ensemble of reality" but engaged in a transfer of "affective power from the image to the body." There are degrees of unqualified intensity at the level of embodied gesture, repetition, and unstructured anticipation amid this type of interaction.

The experience encourages individual engagement with these objects, as well as the exchange of information, objects, and images through virtual spaces. Elements such as "virtual graffiti" and "tracking" further complicate things by encouraging surveillance of and between members as they excavate along the trail. In the end, the complexities that appear around virtual and real space, the authority of human participants, artifacts and objects, and the physical space itself, create competing vertical and horizontal registers for exploring what kinds of circulations, patterns, and displacements of information and meaning can materialize.

The final phase of the project comes in the form of performances, art, and writing projects that draw from connections between Morgan's narrative of the Hollins Community and the information and images recorded from the on-site investigations. Conspicuously absent from either phase of the installation are interfaces and performances that would project these experiences as coherent, representational narratives that "tell" one history or counter-history of the space.

Cathy Caruth has argued that embodied trauma and history are similar phenomena: "trauma is not locatable in the simple violent or original event in an individual's past, but rather in the way that its very unassimilated nature—the way it was precisely not known in the first instance—returns to haunt the survivor later on."[7] Thus, "our recorded histories are symptoms of our desire to recapture the moment we have been unable to escape as a result of our **inability to fully experience it at the time**." There is tremendous potential in this characterization of history and bodily memory. How can we engage with the "traumas" (past and persistent) of a place without relying on a reductive re-colonization through re-telling of counter-experience, with or without the aid of the virtual? If history is indeed a "lost moment that repeats itself" – and not just through narrative but at the level of cognitive and embodied response – there is tremendous potential in imagining new affective environmental tangles (technology, bodies, narratives, objects) that focus on the incipient, mediated aspects of narrative-making. The Hollins Community installation explores the extent to which digital technologies permit a kind of serious play with temporality, spatiality, and embodied affect. The new media interface is typically imagined in terms of its capacities for "access" (seamless or interrupted) and immersion, but there is perhaps even greater potential in the "gaps" that emerge around differing modes of temporality, spatial definition, and bodily anticipation and response within digital environments. In such instances, embodied performance become sites that allow for an experience of alterity as immanent and embodied – indeed, prior to any immersive identification with a particular text or spatiotemporal image. The heterotopic "place" in this project is envisioned not as the re-collection of the layers to existing cultural memory, but as the ever-present threshold of unassimilated histories and traumas and their inscriptions within embodied experience.

Endnotes

[1] Foucault, Michel. "Different Spaces." *Aesthetics, Method, and Epistemology, Essential Works of Foucault* 1954-1984. Trans. Robert Hurley et al. (New York and Paris, 1994), pp.175-85. On space and the semantics of place, see the work of Virginia Polytechnic and State University collaborators, Steve Harrison and Deborah Tatar. "Places: People, Events, Loci. The relation of semantic frames in the construction of place" *Journal of Computer Supported Cooperative Work* (Kluwer Publishers) (in press).

[2] For some examples see the Hollins University website
<http://www.hollins.edu/>

[3] For discussions of the relevance of the "new" in new media See Mark B.N. Hansen's *New Philosophy for New Media* (Cambridge, Mass.: MIT Press, 2004)

[4] Derrida, Jacques. *Memoirs of the Blind: The Self-Portrait and Other Ruins*. Trans. Pascale-Anne Brault and Michael Naas. (Chicago: University of Chicago Press, 1993)

[5] <http://www.tate.org.uk/learning/thamesdig/flash.htm>

[6] Hansen, 207

[7] Cathy Caruth, *Unclaimed Experience: Trauma, Narrative, and History* (Baltimore and London: Johns Hopkins University Press, 1996), p. 4

An Arts Metaverse: Reconstructing the Past
(A Short Review)

Tim Wang and Ulrich Rauch

1. INTRODUCTION

We come here with questions… not answers and we hope that attending and listening and engaging will help us to clarify some of our fuzzy thinking on the relevance and impact of 3D virtual environments on teaching and learning… no… more ambitiously… on the way we perceive, and actively construct the world around us. You can see: We take a fairly subjectivist stance. The power of immersive environments—whether these are provided by film or a virtual reality such as our modeled a First Nations Longhouse, a Greek temple, or virtually skiing down a powder-sugared slope—manifests itself in the crossing of "psychological thresholds" that encourage new perspectives, resulting in a "shift of consciousness" (Hovagimyan). Without becoming too poetic, 3D VL environments may well change teaching, learning, and scholarship. As such we engage in transformative, and if you want, subversive work.

2. WHAT IS IT?

The Ancient Spaces project draws on 3D gaming technologies, the skill of student modelers, and the expertise of faculty to bring Mediterranean (Egypt, Mesopotamia, Greece, and Rome) and North American antiquity alive for teachers, learners and the public at large. Created by a collaborative effort between UBC students, faculty, and staff in 2003, Ancient Spaces is the precursor to an Arts Metaverse and is an attempt to recreate entire civilizations in virtual space through the collaboration of historians, archaeologists, architects and students.

The technology developed enables students to reconstruct the monuments of ancient civilizations in an interactive 3D simulation. Thus far, these reconstructions have included:
A Nisga'a Village (Nass Valley, BC)
Sections of Machu Picchu (Peru)
Hierakonpolis and Deir el-Medina (Egypt)
Acropolis and Agora (Athens)

Together, these reconstructions help to shed light on the core values of material culture in a wide range of global cultures, ranging from Europe to the Near East and the aboriginal civilizations of the Americas. All of these are slated to become complex digital worlds for virtual explorers.

One of the key goals of the project is to produce technology comparable in quality to that which goes into the creation of computer games, but (1) freely available in the public domain and "open-source", so that any academic or member of the public can edit it, and (2) designed for fully educational use that allows students to control content creation.

3. HOW IT BEGAN

2003-2006 Proposals for Funding (note: the shift from "gaming" to self-defined, self-organized VLE)

AN ONLINE ENVIRONMENT FOR THE INTERACTIVE STUDY OF THE ANCIENT WORLD

We propose to enable the student to reconstruct and re-experience the material culture of ancient Greece, Rome and the Near East in a collaborative environment. This project aims to provide a supplementary infrastructure for several core curricula within the department of Classical, Near Eastern and Religious Studies by encouraging student participation and engagement with the ancient world through digital media. A modular approach will allow the foundational design to be re-used in future projects elsewhere in the Faculty.

Ancient Spaces has been student-driven from the beginning, and continually based on open-source software. The idea for a student-built, "massively multiplayer" world based on classical antiquity was put forward by Michael Griffin, then an undergraduate student in UBC's Department of Classical, Near Eastern, and Religious Studies in January 2003, and an early version of the software was written in the open-source library CrystalSpace (focusing on the Palace of Minos at Knossos). In July 2003, a cross-disciplinary group of students drawn from Classics and Computer Science, all with ties to the Faculty of Arts' Instructional Support and Information Technology (Arts ISIT) unit at the University of British Columbia initiated the project idea. The group consisted of three students: Michael Griffin, Dieter Buys, and Jo McFetridge, the co-founders of the project (http://ancient.arts.ubc.ca, see the team page).

Using a "mode" of a gaming platform called Unreal Tournament 2004 these students demonstrated that traditional gaming technology could be put to use to create a realistic and explorable 3D model of the ancient Athenian Acropolis.

Core Goals

1) To generate an infrastructure that enables non-expert users to model key aspects of the ancient world.

2) To educate faculty in the use of these learning objects for creating cinematic and interactive illustrations of key historical events in the western tradition.

3) To engage students in the active use of this technology to complement projects in those same curricula.

4) Using this infrastructure, to create simulations of the social and religious context of daily life in the major centres of the ancient Mediterranean world.

Some examples of the ancient Greek simulations created using the game engine:

http://ancient.arts.ubc.ca/images/movs/1Propylaea.avi
http://ancient.arts.ubc.ca/images/movs/2Propylaea.avi
http://ancient.arts.ubc.ca/images/movs/3Acropolis.avi
http://ancient.arts.ubc.ca/images/movs/4Acropolis.avi
http://ancient.arts.ubc.ca/images/movs/5Acropolis.avi
http://ancient.arts.ubc.ca/images/movs/6Acropolis.avi
http://ancient.arts.ubc.ca/images/movs/7Acropolis.avi

2004

Beginning from this proof of concept, Ancient Spaces, with support from UBC's Teaching & Learning Enhancement Fund (TLEF), piloted the project in a first year classical studies course, Classical Studies 100, at the University of British Columbia. A volunteer group of twenty classical studies students replaced their traditional essay with an immersive 3D reconstruction of the ancient Athenian Agora, including major structures such as the fifth-century Temple of Hephaestus and the Tholos or Council-House, where many crucial decisions of the prototypical democracy were made. The technical elements of the project were simple and "backgrounded" so that the students required little technical expertise. Feedback from students was overwhelmingly positive.

2005

With the support of a larger TLEF of $80,000, Ancient Spaces set out to develop their our own unique and open source technology to allow any university to contribute content, and to "background" the need for technical expertise still further. Ancient Spaces, also, began to considerably expand the range of areas to be modeled from Ancient Athens to Ancient Egypt, British Columbia, and Machu Picchu in Peru.

New technology makes possible the rendering of these famous ancient places into a 3D format with the ability to engage in a virtual tour of them. Moreover, students will eventually be able to interact with these sites by taking an active role in building the Parthenon, or reconstructing the Agora, the pyramids of Egypt, along with virtually experiencing digs at the Lunt (Britain), Stymphalos (Greece), Monte Polizzo (Sicily), the royal tombs at Abydos (Egypt), Hierakonpolis (Egypt), and Tell Acharneh (Syria). Acquiring direct, hands-on knowledge via the virtual world along with a rich visual experience would be of benefit to every student in the department and, once fully developed, of benefit to the community at large. These virtual classrooms would be models of learning, retention and, through virtual visits to these ancient sites, internationalization. In this regard, Ancient Spaces would enable students to reconstruct and re-experience the material cultures of the ancient Mediterranean world in a collaborative environment.

2006
Ancient Spaces: student–driven reconstructions of ancient civilizations

Ancient Spaces enables the undergraduate archaeologist to rebuild the monuments of ancient civilizations in an interactive 3D simulation. This new approach to "post-constructive learning" in archaeology uses the creation of immersive computer simulations which draw upon e.g. archaeological excavations led by UBC instructors or existing records of excavations, to create an immersive process that allows students to engage with the recreation of physical artifacts, but also the recreation and understanding of the social and cultural environments in which artifacts became formed. The simulations are then reviewed in the academic community, and subsequently showcased and shared (via the Internet) as original undergraduate research. The project was initially conceived and led by students, with students leading the development of the interactive 3-D simulations. Academic support for the project came originally from the Department of Classical, Near Eastern and Religious studies, but has widened to include the First Nations Studies Program and the Department of Art History, Visual Art and Theory. Each class complemented by Ancient Spaces also produces a learning environment for the next cohort of students, affording undergraduates a leading role in experiencing new postconstructivist pedagogical approaches by interacting equally with subject matter, with peers and with supervisors.

4. WHY ARE WE DOING THIS?

The project was designed initially as an aid for teaching and learning in departmental curricula by presenting students with an alternative way of experiencing the ancient world, and allowing them to participate actively in its reconstruction. Ancient Spaces provides students with the opportunity to engage in experiential learning. The project also aids greatly in the acquisition of Information Technology literacy among Arts students. The project also seeks to provide digital forums for peer-to-peer and teacher-to-student academic discourse, and to promote the development of multiple learning channels for Humanities students. In this spirit, Ancient Spaces aims to enable students and non-academics to jointly become researchers in creating and sharing knowledge beyond the walls of the university. We hope to create a simple interface and technology for sharing, qualifying, and evaluating interactive three-dimensional content.

How?

Students would build this interactive world themselves, element by element, individually or in teams, by conducting academic research into the form and function of an individual ancient building, event, artifact or ritual, and using pre-made 3D objects to reconstruct the ancient element in 3D. The result is a modular, digital world that can be used and reused to complement and enhance existing course curricula and as a digital forum (through a wiki) for students to debate particular archaeological, religious, or historical reconstructions.

5. SOME OUTCOMES

Together, these reconstructions help to shed light on the core values of material culture in a wide range of global cultures, ranging from Europe to the Near East and the aboriginal civilizations of the Americas.

In practice, each student focuses on the creative process of gathering research to generate an accurate photo-realistic three-dimensional image from a twodimensional site plan or written report. The environment created, accompanied by an oral or written report on the decisions made in converting descriptive data to a fully explorable space, serves as a term project. Combining projects, each cohort of students generates a large and compelling, but static environment. Using the Ancient Spaces Editor, the same environment can be transformed and different theories of reconstruction explored, demonstrating the dynamic nature of archaeological knowledge. In application in First Nations Studies (INDS 530B), this approach also allows student creations to be shared with Elders and knowledge-holders. Elders from the Nass Valley did evaluate at the end of the semester further develop student work.

"The potential of the course was just phenomenal," said Nyce {a student}, who is also the president of the Nisga'a House of Wisdom, a non-profit university college in the village of New Aiyansh, BC, "aboriginal thought is not common to everybody's common knowledge, and contextualized aboriginal thought is even more remote, unless you spend time in the community."

As well as its direct application in these courses, the Ancient Spaces technology is being used in research applications within the Faculty of Arts. For example, a novel archaeological site in Egypt, currently in the process of excavation, is being directly recorded in the high-resolution 3D environment of Ancient Spaces. A Canada Research Chair-funded project in the Department of Psychology is making use of the same 3D approach to create content for a research study in the human perception of dimensionality.

Benefits student-centred active learning: In place of the in-class, slide-show approach to the study of antiquity, art history, and archaeology, the Ancient Spaces 3D modelling program asks students to engage in investigative practises, work with field data, interpret forensic evidence, and weigh competing theories. Students gain a better understanding of the ways in which a lost culture's architectural choices can shed light on its social dynamics and core values.

Teaching the conflicts: While there will always be a need in archaeology for museum-quality site reconstructions produced with the help of expensive scanners, the Ancient Spaces approach to the production of good quality models makes it easy for students to demonstrate their knowledge of varying theories by producing different replicas of the same site reflecting interpretive conflicts in the field.

Student engagement: Change the nature of the study of society and culture in the Humanities by putting students inside their subjects and giving them the freedom to make their own discoveries based on an interactive model.

Perspective taking: With the "inexplicable interaction" between mind and matter manifest in the interaction between self inside and outside a 3D VR environment the potential to challenge social and cultural explanatory models that are based on classical subject-object distinctions… is gratifying. The adoption of digital technology in the Arts, Humanities and Social Sciences encourages interactive participation in immersive experiences and thereby enables questions of representation, perception, and cognition in relation to the production of meaning.

A VR simulation in this context might offer profound insights into "how the world is… conceptually organized and integrated" for indigenous cultures in that by encouraging an immersive experience it might obviate what McPherson and Rabb refer to ("Indigeneity") as the "outside-view predicate."

4. QUESTIONS

We ask: will a Metaverse, and our experience of living in two worlds help us to improve our own understanding on how "particular institutions, media, texts, discourses and disciplines are inhabited, haunted, even constituted by what they cannot tolerate, by what they cannot acknowledge, by what is alien, external, contaminatory" (Nicolas Royle, *After Derrida*).

And are we passing a new threshold of consciousness by expanding our way of thinking and by accepting of being in two (multiple) worlds, or are we falling back in yet another dialectic of enlightenment because we fail to acknowledge how virtual world and live-world are but one?

5. A FINAL SHOWCASE OF THE ARTS METAVERSE

http://artsmetaverse.arts.ubc.ca/media/artsmetaverse.avi

Acknowledgments

We want to acknowledge the contributors of a variety of sources to this synopsis:

Michael Griffin, Ph.D student, Classics, Oxford University
Jodey Castricano, Associate Professor Critical Studies, University of British Columbia Okanagan
Marilyn M. Lombardi, Director, Senior IT Strategist & ISIS Senior Research Scholar, Duke University
Liang Shao, Graphic Designer, University of British Columbia
Bryan Zandberg , Writer for UBC's artsBeat

Finding Humanity within the Machine: Large Motor Movement Computer Interfacing as an Artistic Mindbody Integrative Practice

John Toenjes and David Marchant

Preface

Our original interface, in all human endeavor, is the moving body and its senses. All extensions we may create digitally are and must be based in our primary sensuous being. Thus, how we ARE and how we move have everything to do with what sort of interfaces we will make.

Demonstration of Performance Environment

We are a collaborative team that creates computer-mediated interactive music and dance performances, which utilize motion tracking systems to create bodily immersive music-making environments. Within this environment we improvise music and visual art through full-bodied dance movement. Our performance aesthetic necessitates a computer interface within which one physically "dwells." This spatial/sonic/visual environment is an extraordinary interface in which the computer mediates the relationship between a person's movement and the resulting music. This makes for a rather magical theatrical experience, and allows us to achieve a unique integration of music and dance, where the music is sounded as a direct result of the movement, and movement aesthetic is conditioned by the act of "being" the music. Additionally, video effects are responsive to movement, which further influences and inspires movement choices and qualities. Thus, our interface becomes a medium through which we can unify the sound, movement, and visual arts.

Integrations: Art, Man, Machines and Environments

Integration of Artistic Mediums

Historically, in traditional dance performances, the choreography, music and scenic elements occur simultaneously, yet remain fundamentally separable. Within our interface, these are simultaneously created of and by each other as a true whole. And while we initially set out to integrate music and dance, what we unexpectedly found while creating and playing the instrument we call *Leonardo's Chimes*, is that this interactive environment is also integrative for the individual who plays it. We discovered an extraordinary self-movement feedback experience that fosters holistic integration of mental and physical faculties, developing kinesthetic, spatial, and musical aptitudes.

Integration of the Person in the Environment

One of the most curious effects we experience is that moving in this environment has the effect of making one feel more "whole," more unified in thought and action. We are intrigued with the notion of making systems that improve our integration as human beings. If the interaction between man and machine brings about the betterment of the man, is the machine even more wholly part of our living humanity?

People are used to interfacing with the environment by taking in information with the eyes and ears, and effecting change with the face and hands [see Figure 1]. Notice in fig. 1 how prominent hands and face are in proportion of cortex motor units. The hand is itself larger than the entire remainder of the lower body. One may experience a radical shift when beginning to interface with the world with a more balanced and whole use of the "bodymind".[1] This is the special domain of the dancer, and is why, typically, computers and dancers may not tend to mix. A large-motor-movement interface demands integrative development of the bodily aspect of our mind. New perceptual doors open with such a shift in perspective.

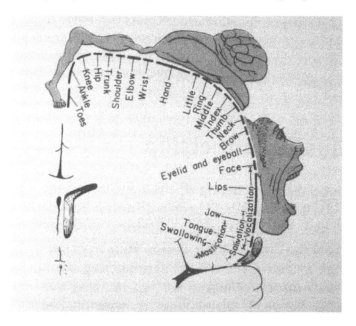

Figure 1: A.C. Guyton: Features are distorted in relative size to represent proportion of cortex motor units dedicated to each body part. Hands and face are most prominently represented in cortex.

A common idea in the debate about the benefits and costs of computers in human life is that the use of computers is likely to cause bodily atrophy, or a split in the relationship between a person's intellectual and physical sides. Imagine the stereotypical computer geek huddling over his terminal, or the child spending hours collapsed on the couch playing video games, eyes pulled inward and down toward the TV and hands tightly gripping the controller. Our interface more ideally addresses three issues in psychophysical well-being. First it discourages isolation of the mental from the physical, by requiring full-bodied movement to interface with the computer system. Second, it encourages full range physical extension outward into space instead of only forward, narrowly focused contraction. And third, movement occurs within a multi-sensory, interactively responsive environment, which provides feedback to the player to help integrate thought and action as a whole.

Feedback Loops: A Mechanism for Synaesthetic-Sensory Integration

This instrument creates and amplifies a "synaesthetic-sensory" interactive response field. Because of the spatial arrangement of the instrument's active areas, the player is encouraged to move with full range of motion to activate sonic and visual responses. When the player hears and sees the results of his movement, he experiences a feedback for the shape, location, timing and quality of the movement. This feedback "tunes" the player's spatial, proprioceptive, and kinesthetic awareness. Furthermore, the player is increasingly inspired to "dance to the music"—dance to *his* music—which generates yet more sound, inspiring the player to move in response. Thus, the experience for the mover is a "feedback loop" in which his movement makes music that inspires him to move in ever-more creative ways to make more sound, and so on…

Skilled, full range motion, that is, "dancing," constitutes the instrument's virtuosic technique. Once basic bodymind understanding of the cause/effect relationship with the response field is established, the feedback loop encourages the player to become more skilled in movement, spatially, temporally and qualitatively.

If making music is the goal, everything we do sonically in *Leonardo's Chimes* could be done with fingertips on laptops. So why bother to make it take up so much space? The reason is the importance of involving more of the human body. Most modern technology is miniaturized, from an assumption it is preferable to minimize movement activities. We made the computer take up an enormous amount of space to require more movement. Involving more of the bodymind in the experience leads to a more enveloping feedback response and imparts an organic quality to the music that seems lacking in much contemporary computer music.

A More Integrative Interface

Through our artistic and aesthetic explorations we have come to realize that this interface has implications beyond the theatrical, into the realms of physical transcendence, integrative therapies, and the experience and contemplation of both the real and the philosophical relationship of "man" to "machine."

Innerspace and Interface
The feedback loops are a subjective experience that constitute the "innerspace" of this interface. When we began to feel this sensory feedback, we noticed that previous intellectual distinctions between movement/sound, self/environment, man/computer began to blur in our experience, replaced by one extraordinary, melded sensation. Thus, one might argue that this interface not only connects previously disparate parts, but also expands the "innerspace" *into* the "outerspace," eliminating separations between person, computer, and the surrounding environment.

In contrast to present definitions of "interface" which suggest the "connecting" or communicating between separate things,[2] we are beginning to prefer the use of the word "medium," because it suggests communion rather than mere connectivity.[3] Using this medium elicits a sensation that one's innerspace is integrated with the outerspace. This sensation has a profound effect upon the very notion of the computer and its role in the human experience.

The Man in the Machine and the Machine in the Man
The integrative nature of this experience begins to blur philosophical and lived distinctions between "man" and "machine." By giving the computer "sight" via cameras, and "thought" via complex parameters for responses, the player begins to feel as though he is "relating" with an entity, as opposed to "using a tool." There develops a sensation of moving in a liminal space, across an increasingly porous membrane of mind and machine, somewhere in between the muscles and the microchips. One begins to feel like they are simultaneously "in" the machine, and that the machine is "in" them. It is a dance—an exchange of sensation and mutual response.

Other Applications of Such Environments

Creativity Development
We invented *Leonardo's Chimes* as an instrument for performance art. It is a medium in which one's imagination is allowed to stretch and experience sensory immersion and interplay in a "super-elastic, magic plastic" world of motion, sight and sound. As a work of theatre, we made the clear decision that this environment should have some of the normal cause/effect relationship of "real world" environments. Otherwise, it is too illogical and confusing. But we also chose ways of bending the rules along certain parameters to make it more intriguing. This sensation of stretching the constraints of physical reality has an uncanny effect on the mind, allowing it to suspend old assumptions and seek new pathways, ideas and solutions. Playing this instrument has been invigorating

to us, not only for our art, but also for our imaginative capacity in general.

We believe this instrument's capacity to inspire creativity could be extended beyond its original artistic intention, because it encourages one's imagination to play outside the normal boundaries of cause and effect. Creativity is the ability to suspend habitual assumptions about how things work to reveal new options, allowing previously unrelated ideas to interact in novel ways. A person's neuro-motor system has habits about how to think about the world. These well-worn pathways are conditioned by the physical world, where cause and effect is relatively predictable. In a computer-mediated environment, some of the world's usual rules can be "bent" or broken (e.g. the ability to pass your hand through the virtual drum one is beating or hearing a beautiful sound that is improbably emanating from your body in motion). This allows the mind of the player or audience member to stretch into new domains of possibility.

Since the brain is working outside its normal frame, one feels the sensation of possibility as a generalized perspective, applicable to other areas where imagination, creativity and problem solving are useful. Such environments may have profound effect on the neurological level, increasing flexibility, both physically and mentally; the mindbody is the real "super-elastic, magic plastic."

Integrative Body Therapies/Physical Education & Rehabilitation

This system offers many possible educational and re-educational applications for developing basic sensory-motor skills, and use in integrative body therapies. In many approaches to rehabilitation and physical education, one of the central issues is cultivating sensory awareness and perceptual accuracy. In particular, proprioceptive and kinesthetic sensitivities must be developed for proper self-perception in time/space movement skills. Because systems like our interface provide such extraordinary feedback about location and timing of movement to the player, it gradually increases awareness of the body in relationship to space and time, fundamental to all movement skill. We suggest that this sort of system could be valuable to the improvement of people's physical coordination. This nicely counters traditional notions of computers contributing to an increasing dissociation of our minds from our bodies. This computer interface actually facilitates integration of one's faculties rather than isolating them. By giving the person a higher order end goal, that is, creating beautiful sound, all lower level faculties are organized into whole thought/actions.

Topics for Discussion/Questions/Further Implications

The experience of playing this instrument not only allowed us to break previous assumptions about the "parts" of a dance performance (dance, music, visual scenery) and make unique, intriguing art, but also opened many wider questions. Following are some additional notes and philosophical musings on ideas about interface, systems, and integration (of machines, of man and of both) toward new, larger wholes.

What is interface?

Interface is defined as "a point where two systems meet and interact,"[4] and "a device or program enabling a user to communicate with a computer." But both of these definitions maintain the dualistic sense of fundamentally separate things merely "interacting."

After playing the instrument we began to experience sensations that challenge dualistic descriptions of the separateness of things and opened philosophical sounding questions such as: "Where am I? Am I only here, where my body is, or am I also there, in the computer? Where is the computer? Is it only there, on the table, or is it also here, inside of me in my mind, my body, my action and response? What is the computer? Is it only a thing I use, or is it a more integral extension of myself?"

Is the word "interface" the ideal term for this or any system? Why not "medium"?

Medium is defined as: 1. An intervening substance through which impressions are conveyed to the senses or a force acts on objects at a distance. 2. The substance in which an organism lives or is cultured. 3. The material used by artist, composer, writer. 4. Storage method for digital data.[5]

Ours is a "spatial, sonic, photonic, kinesthetic/neurological medium"— a synaesthetic medium through which a person kinesthetically experiences Self. This instrument is a unique medium for us to relate to each other through space, time, touch, sound, and sight.

Why does playing this instrument feel "unifying," or "integrating" for the player?

When playing this instrument, I don't feel the usual physical boundaries, compared with, say, striking a drum. Tactility might typically give one a sensation of separateness from the thing being touched. But because I can't feel a surface to touch, I begin to sense the space proprioceptively; place becomes an internal sensation—this brings the outer to the innerspace (arguably this is always true, even with tactile touch, but perhaps becomes more curious or obvious a sensation when it happens in the absence of tactile feedback). Thus, the only way to feel a location or time in this space is to feel *me* more accurately—an idea which we call "self-referential space."

Self-referential space and time

Time and space are fundamentally two ways of describing the experience of measuring movement. The original reference point for any measurement of space or time is the human body's own dimensions and the kinesthetic/proprioceptive meaning of "here vs. there" and the duration of change between two perspectives. Without movement (or perception of change), there would be no experienced time or space.

Playing the instrument tunes kinesthesia and time/space skill. Time and space are bodily sensations first, and intellectual ideas second. So one implication of playing this instrument is that it tunes this "original reference" for space and time—the Self—and therefore is an intriguing tool for improving one's accuracy of time/space movement skills.

What does extension and large movement have to do with the integrative experience?

Higher-Order Integrative Tasks

Movement control skills involved in playing this instrument are not done merely for some arbitrary objective of physical control (an artistic flaw to which dance techniques occasionally fall prey). Rather, such control is also the necessary means to play music and make video art, which becomes a higher-order organizing objective for one's movement skills. One's bodymind not only manages balance, but is also engaged in evaluation and execution of a spontaneous musical experience, integrating areas of thought and action into one problem-solving activity. Perhaps because the task of balancing on one foot and stretching in two directions is not the end goal, but merely part of an overall solution, balance and control become easier because they are employed toward higher order goals. The basketball player leaps very high indeed, but not because he is focused on the leap; his larger objective (to put the ball through the hoop) naturally organizes all necessary lower order skills to achieve a higher end.

Although we have not yet experimented with different spatial arrangements of active areas, it occurs to us that this system offers flexibility of configuration, which would allow us to design it to encourage movement for a variety of shape, line, or physical control goals.

What exactly do we mean when we say that playing this instrument integrates the man, and the "man in the machine"?

Innerspace experience affects the Interface, which in turn affects the Innerspace, which affects the Interface… they both are mutually influential aspects of a whole.

There is a phenomenal[6] sensation when I am in this environment that is as if my mind is no longer only in my head, but also extends into my body, and then spreading out, surrounding me in space. It is as if my brain is a pool in which I am immersed, and yet at the same time, the computer-mediated field is also immersed "inside" me, via my senses and response/actions to and in that field.

Arthur Koestler coined the term "holons" to refer to that which is whole in one context and yet is also only part of a wider whole in a larger frame or system (Wilber, 36). Ken Wilber says, "In any developmental sequence, what is whole at one stage becomes merely a part of a larger whole at the next stage. A [whole] letter is part of a whole word, which is part of a whole sentence, which is part of a whole paragraph, and so on." He speaks of "…expanding links…[and] an increase in unity and wider identities…" as a way to illuminate the meaning of Koestler's concept of "holarchies" (Wilber, 36). Holarchies are like hierarchies, but holistically expanding. When a holarchy expands to higher levels, it extends (transcends) AND includes all previous levels, analogized by a set of nested Russian dolls. In the Human Being, the holarchy expands from elements to compounds, to proteins, to organelles, to cells, to organs, to organ systems, to brain, to mind, and so on. In the computer, it unfolds from elements to compounds, to materials, to chips and circuits, to hard drives, to operating systems, and so on. "Interface" is just a way of trying to describe a fuzzy zone where the holarchies of computer and person overlap and begin to merge into still larger holarchies, now with the machine and the man operating as new, larger holon (whole, but still only part of yet larger systems, such as the internet, which is whole, but still only part of a wider community of people, and so on).

So one more time... what is interface?

When interface is properly functioning, it does not merely connect entities at a boundary; it eliminates boundaries/separations between entities and enfolds entities into one another. The purpose of a good interface is to eliminate separations and make a seamless, unified sensation of thought transforming into action/response unifying us with the world. In this sense, one's mind is extended by body, which is further extended into and through computers to "locations" in space (and virtual spaces) of an ever-extending meaning of "world."

Interface is really more of a concept than a thing. So our ideas about interface—the way we describe and think about it—define its capabilities and its limitations. If we only think of interface as a way of connecting separate things, then it implies that the things (person, machine, environment) are still fundamentally separate, but connected. But perhaps the real goal of interface is to unify separate things into new, whole, integral systems.

Endnotes

1 "Bodymind" is one term among several variations that attempts to correct dualistic notions that the body and mind are fundamentally separate but connected. It is an attempt to speak more properly of the whole, integrated nature of a person, mental and physical aspects functioning as a single continuum.

2 Definition: Interface is "a point where two entities meet and interact."

3 Definition: Medium is 1. An intervening substance through which impressions are conveyed to the senses or a force acts on objects at a distance. 2. The substance in which an organism lives or is cultured. 3. The material used by artist, composer, writer. 4. Storage method for digital data.

(All definitions are from the *Oxford American Dictionary*, as implemented in the "Dashboard" "widget" in the Macintosh OSX interface.)

4 *Oxford American Dictionary*

5 *Oxford American Dictionary*

6 I use the term "phenomenal" here in reference to "phenomenology," a philosophical investigation of the subjectivity of lived experience, proposed by Edmund Husserl and Maurice Merleau-Ponty. A central intent of their philosophy was to attempt to create a "…rigorous science…" but one that "…offers an account of space, time and the world as we 'live' them…to give a direct description of our experience as it is, without taking account of its psychological origin and the causal explanations which the scientist, the historian or the sociologist may be able to provide." (Merleau-Ponty, vii).

References

Guyton, A.C. 1971. *Basic Human Physiology: Normal Function and Mechanisms of Disease.* Philadelphia: W.B. Saunders Co.

Husserl, Edmund. 1960. *Cartesian Meditations: An Introduction to Phenomenology.* Translated by Dorian Cairns. The Hague: Martinus Nijhoff Publishers.

Koestler, Arthur. 1976. *The Ghost in the Machine.* New York: Random House.

Merleau-Ponty, Maurice. 1962. *Phenomenology of Perception.* Translated by Colin Smith. London: Routledge.

Wilber, Ken. 2001. *The Eye of Spirit: An Integral Vision for a World Gone Slightly Mad.* Boston & London: Shambhala.

Oxford American Dictionary, as implemented in the "dashboard widget" of the Apple Macintosh OSX interface.

Ludic Depths: Games, Narratives, Platforms

Complex and sometimes contradictory notions of narrative play out in hardware and software design, games structures, and historical modeling and pedagogy.

Ian Bogost and Nick Montfort
"New Media as Material Constraint: An Introduction to Platform Studies"

Patricia Seed
"Looking Back: A Decade of Using Games to Teach History, 1996-2006"

Noah Wardrip-Fruin
"Internal Processes and Interface Effects: Three Relationships in Play"

New Media as Material Constraint: An Introduction to Platform Studies

Ian Bogost and Nick Montfort

ABSTRACT: We introduce platform studies, a family of approaches to digital media. In platform studies, close consideration is given to the detailed technical workings of computing systems. This allows the connections between platform technologies and creative production to be investigated. Two short studies of the Atari VCS (2600) and the Nintendo Wii show how close consideration of this sort can inform our understanding of the history, present, and future of new media.

Platforms have been around for decades, right under our video games and digital art. Those studying new media are starting to explore the low level of code to learn more about how computers are used in culture, but there have been few attempts to go even deeper, to the metal — to look at the base hardware and software systems that provide the foundation for computational expression. Platform studies is such an attempt, investigating the relationships between the hardware and software design of standardized computing systems — platforms — and the creative works produced on those platforms.

INTRODUCING PLATFORMS

The hardware and software framework that supports other programs is referred to in computing as a platform. A platform in its purest form is an abstraction, simply a standard or specification. To be used by people and to take part in our culture directly, a platform must manifest itself materially. This can be done in the chips, casings, peripherals, and other components that make up the hardware of a physical computer system. A platform may also include an operating system. It is often useful to see a programming language or environment on top of an operating system as a platform, too. Whatever the programmer takes for granted when developing, and whatever, from another side, the user is required to have working in order to use particular software, is the platform. In general, platforms are layered — from hardware through operating system and into other software layers — and they relate to modular components, such as optional controllers and cards. Studies in computer science and engineering have addressed the question of how platforms are best developed. Studies in new media have addressed the cultural relevance of particular software that runs on platforms. But little work has been done on how the hardware and software of platforms influences, facilitates, or constrains particular forms of computational expression.

By choosing a platform, new media creators simplify development and delivery in many ways. Their work is supported and constrained by what this platform can do. Sometimes the influence is obvious: A monochrome platform can't display color, a video game console without a keyboard can't accept typed input. But there are many more subtle ways that platforms interact with creative production, due to the idioms of programming that a language supports or due to transistor-level decisions made in video and audio hardware. In addition to allowing certain developments and precluding others, platforms also encourage and discourage different sorts of expressive new media work with much more subtlety. In drawing raster graphics, the difference between setting up one scan line at a time, having video RAM with support for tiles and sprites, or having a native 3D model can end up being much more important than a few numbers representing resolution and color depth.

Particular platform studies may emphasize different technical or cultural aspects and draw on different critical and theoretical approaches, but to deal deeply with platforms and new media, these sorts of studies will all have to be technically rigorous. The detailed analysis of hardware and code can connect to the experience of developers who created software for a platform and users who interacted with and will interact with programs on that platform. Only the deep investigation of computing systems will reveal the interactions between these systems and creativity, design, expression, and culture.

New media studies, focusing on artifacts, games, and works of digital art and literature, have been undertaken on many different levels. The ones we describe next have been discussed in the context of a specific video game before, but here we briefly explain how they are relevant to digital media studies overall.[1] This provides some context for our focus on the lowest level, that of *platform*.

Reception/operation is the level that includes reception aesthetics, reader response theory, psychoanalysis, desensitization to violence studies, and empirical studies of interaction and play. Only interactive media are explicitly operated, but all sorts of media are received and understood, so insights from other fields can often be usefully adapted to digital media at this level.

Interface studies include the whole discipline of human-computer interface, comparative studies of user interface done by humanists and literary critics, and approaches from visual studies, film theory, and art history. Remediation concerns itself with interface, although reception and operation are concerns of that approach, too. This is not unusual. Many new media studies span multiple levels, but there is often a focus on one.

Form/function is the main concern of cybertext studies and of much of the work in game studies and ludology. Narratology, previously used to understand literature and cinema, is an approach that deals with form and function and which has been applied to new media as well. Because these approaches deal with the same level, it is at least meaningful to imagine a narratology/ludology debate — whether or not any such debate has occurred — while it makes much less sense to think about a psychoanalysis/ludology debate or a remediation/narratology debate.

Code is a level that has only recently been explored by those investigating new media. Code studies, software studies, and code aesthetics are not yet widespread, but a few interesting books and panels dealing with the code level signal an increasing interest in the way creative work is programmed and understood by programmers. The discipline of software engineering is a related field that concerns itself with the code level as well as with organizational and individual capabilities for software development.

Platform is the abstraction level beneath code, a level which has not yet been systematically studied. If code studies are new media's analogue to software engineering and computer programming, platform studies are the humanistic parallel of computing systems and computer architecture, connecting the fundamentals of new media work to the cultures in which they were produced and the cultures in which coding, forms, interfaces, and eventual use are layered upon them.

Our focus here in on this platform level. We hope that studies at this level will help to fill in our overall understanding of new media and will benefit the humanistic exploration of computing. We also want to emphasize that we see all of these levels, not just the top one, as being situated in culture, society, economy, and history. Because of this, we seek to describe how platforms have come about as well as how they influence further cultural production. This awareness of contexts informs our approach to platform studies, just as it has informed the best new media studies at other levels in the past.

Next, we discuss three examples from different eras of computing to explain the general relevance of platform studies in new media. Our examples include two video game consoles and one general-purpose personal computer system. They are the 1977 Atari Video Computer System (later called the Atari 2600), the 1991 Multimedia Personal Computer, and the 2006 Nintendo Wii.

1977: THE ATARI VCS (2600)

The Atari VCS (renamed the Atari 2600 in 1982) was the first successful cartridge-based video game console. We describe the elements of the Atari VCS: The 6507 processor, its Television Interface Adapter (TIA), its interchangeable ROM cartridges, and a variety of 8-pin controllers.[2] These can help why VCS games imitated some aspects of arcade games while they left other aspects aside. We conclude this section with a detailed discussion of how sprite graphics worked on the VCS and how this influenced the development of cartridges.

The Atari VCS appeared at a time when the vast majority of video games were played in bars, lounges, and arcades. Today, the arcade cabinet is a rare sight, but in their heyday coin-operated games generated more income than today's home console-dominated market.[3] At a time when coin-ops ruled the market, part of the appeal of the home console system was its promise to tap a new market of kids and families.

That year Atari, eyeing the home market for video games, also designed a home *Pong* and arranged for Sears to sell it exclusively — which they did, moving 150,000 units in 1975.[4] Atari's triumph was short-lived, however. In 1976 General Instrument released its $5 AY-3-8500, a "PONG-on-a-chip" that also contained simple shooting games. It, along with other cheap processors, allowed even companies without much electronics experience to bring *Pong*-like games to market. They did — there were 75 available by the end of 1996, "being produced in the millions for a few dollars apiece." Even if Atari had cornered the market for home *Pong*, having that system in a home wouldn't have done anything to lead to future sales. How many *Pong*s could one house have needed? Atari looked, instead, to model some features of the nascent personal computer market with a home console that used interchangeable cartridges, allowing the system to play many games. There would be an important difference from home computing, as Atari saw it, though: All of the cartridges for the system would be made by one company.

The tremendous success of *Pong* and the home *Pong* consoles suggested that Atari produce a machine capable of playing *Pong*-like games. The additional success of *Tank* by Kee Games (a pseudo-competitor that Atari CEO Bushnell created to give the sense of an industry) suggested similar design possibilities for what would become the VCS. *Tank* featured two player objects, each controllable by a separate human player, and projectiles that bounced off walls — a computational model almost identical to *Pong*, and one that would become the inspiration for *Combat*, the title that was included with the original VCS package. These simple, existing elements would be the basis for the console's capabilities.

On the other hand, previous attempts at home machines that used interchangeable cartridges, such as the Magnavox Odyssey and the Fairchild VES/Channel F, suggested the potential benefits and risks for such a system. Released in 1972, the Odyssey played twelve games, but required players to attach plastic overlays to the screen in lieu of a computer graphics background. The machine had no memory or processor, and the experience of playing the Odyssey was certainly that of a video game, but perhaps too simplified, even for the time, and reminiscent of board game play. Even though Magnavox sold $22 million worth of Odysseys by 1975, the product posted losses of $60 million due to distribution and marketing problems — many customers thought they needed a Magnavox TV to play it.[5] Fairchild's Video Entertainment System, released in 1976, was the first programmable, interchangeable cartridge system, with an onboard processor and RAM. (The system had a rapid name change when Atari's VCS was released, and is better-known as the Fairchild Channel F.) Even before Fairchild's system was market tested, Warner Communication purchased Atari in 1976, largely based on the commercial promise of an interchangeable, programmable home console. It was this acquisition that provided the capital Atari needed to bring the VCS to market.

The Design of the VCS

The engineers developing the VCS needed to take into account both of these goals — imitation of known successes and versatility — as they designed the circuitry for a special purpose microcomputer for video games. Material factors certainly influenced the design, most notably the high cost of hardware components. The Fairchild system used the complex Fairchild F8 CPU, a specialty processor created by future Intel founder Robert Noyce. In 1975, MOS Technology released a new processor, the 6502. The chip was the cheapest CPU on the market at the time by far, and it was also faster than competing chips like the Motorola 6800 and the Intel 8080. The 6502's low cost and high performance made it an immensely popular processor for more than a decade. The chip drove the Apple I and Apple II, the Commodore PET and Commodore 64, the Atari 400 and 800 home computers, and the Nintendo Entertainment System (NES).

Steve Mayer and Ron Milner were chiefs at Cyan Engineering, a consulting firm Atari had purchased in 1975. They selected a chip for the VCS project that was very similar to the 6502, but stripped-down and even less costly. The two used the 6507, which came in a cheaper package with only 13 address lines, used to designate which byte in memory will be read or written. This was reduced from the 6502's 16 address lines. So while the low-cost 6502 could address $2^{16} = 64KB$ of memory, the even lower-cost 6507 was only capable of addressing $2^{13} = 8KB$. But the memory that was to be addressed was on cartridge ROMs, and, again in the interests of economy, the VCS was designed with a cartridge interface had one fewer line than the processor would support — limiting access to 4KB of cartridge ROM at once. Bill Gates may have thought that 640KB should be enough for anybody; Mayer and Milner figured that 4KB would do. The 6507 was available for less than $25; similarly capable Intel and Motorola chips went for $200. Using the chip enabled the VCS to have a very low initial retail price of $199 — just above the console's manufacturing cost. This was an unusual move by Atari, but the company was counting on profiting from cartridge sales.

While important, the 6507 CPU was only one component — the VCS still needed additional silicon for memory, input, graphics, and sound. For sound and graphics, the VCS was to use a custom chip, the Television Interface Adapter (TIA). Joe Decuir and Jay Miner designed the TIA, codenamed Stella — a name also used for the VCS itself. Of course, the two sought to simplify the hardware design as much as possible, reducing its complexity and cost.[6] Input from the two player controls and the console switches were managed by an interface called RIOT. The VCS also sported 128 bytes of RAM — not enough to store this ASCII-encoded paragraph. Contemporary home computers usually have more than a thousand times as much memory, but this wasn't an unusual amount for a video game system of the late 1970s. The VCS, like other cartridge-based systems, ran programs without loading them into RAM. The VCS's 128-byte memory was twice as large as the VES/Channel F's standard RAM; the later NES has only 16 times as much RAM, 2KB.

Drawing the Screen

The bare-bones nature of the TIA makes seemingly basic tasks like drawing the game's screen complex. An ordinary television picture of the late 1970s and early 1980s was displayed by a cathode ray tube (CRT). The CRT fires patterns of electrons at a phosphorescent screen, which glows to create the visible picture. The screen image is not drawn all at once, but in individual scan lines, each of which is created as the electron gun passes from side to side across the screen. After each line, the beam turns off and the gun resets its position at the start of the next line. It continues this process for as many scanlines as the TV image requires. Then it turns off again and resets its position at the start of the screen.

Modern computer systems offer a frame buffer, a space in memory to which the programmer can write graphics information for one entire screen draw. This facility was even provided by many systems of the late 1970s — including the Fairchild VES/Channel F that preceded the VCS. In a frame buffered graphics system, the computer's video hardware automates the process of translating the information in memory for display on the screen, also managing graphical administrativa such as screen synchronization.

The VCS does not provide such services for graphics rendering. The machine isn't even equipped with enough memory to store an entire screen's worth of data in a frame buffer. The VCS offers 128 bytes of RAM total — not even enough to store one 8-bit color value for every *line* of the VCS's 191-line visible display, let alone for multiple elements per line such as individual pixels, or, at a higher level, backgrounds, players, and missiles. Additionally, the interface between the processor and the television is not automated as it is in a frame buffered graphics system. Instead, the VCS programmer must draw the screen manually, synchronizing the 6507 processor instructions to the television's electron gun via the TIA. The programmer has a small amount of time to change the TIA settings via its numerous addressable registers when the electron beam resets to draw a new line (this period is called Horizontal Blank), or a new screen (this period is called Vertical Blank). However, the programmer must also manually instruct the TIA to initiate or wait for the horizontal and vertical blanks themselves, which involves keeping track of how much time the instructions take to execute on a single line, between lines, and between frames. Programming the VCS, then, effectively means drawing every line of the television display individually, making decisions about how to change the display on a line-by-line basis rather than a screen-by-screen basis.

Sprites

Many of the common techniques on the VCS are rooted in the manual nature of graphics programming for the device. Despite its simplicity, combinations of hardware and software techniques have produced a wide variety of visual, audio, and gameplay effects in many hundreds of games created in the three decades since the console's release. Rather than try to describe all of these in cursory detail, we focus on one aspect of VCS games: sprites.

In computer graphics, a sprite is a two-dimensional image composited onto a two- or three-dimensional scene. The VCS was designed to support two sprites, each a single byte in size, set via two memory-mapped registers on the TIA (named GRP0 and GRP1, respectively). The influence of *Pong* and *Tank* can be seen clearly here. Such games feature two opponents, each controlled by a human player. The VCS provided a facility for a single-pixel *Pong*-like ball, single-pixel *Tank*-like missiles, and the player sprites that were common to both games.

Sprite Graphics

When the programmer stores a value in the GRP0 or GRP1 register, the TIA displays that 8-bit pattern on-screen. A VCS sprite is thus always 8-bits wide, although the TIA provided a few ways of modifying the appearance of sprites on screen.

```
Bit        7 6 5 4 3 2 1 0         Sprite:
Line 0     0 0 1 1 1 1 0 0          XXXX
Line 1     0 1 1 1 1 1 1 0         XXXXXX
Line 2     0 1 0 1 1 0 1 0         X XX X
Line 3     1 1 1 1 1 1 1 1        XXXXXXXX
Line 4     1 0 1 0 0 1 0 1        X X  X X
Line 5     1 0 0 1 1 0 0 1        X  XX  X
Line 6     0 1 0 1 1 0 1 0         X XX X
Line 7     0 1 0 1 1 0 1 0         X XX X
Line 8     0 1 0 0 0 0 1 0         X    X
```

Figure 1: VCS sprite pattern from *Space Invaders*.

Figure 1 shows the pattern for a sprite — a 2D image, but one that is drawn, like everything on the VCS, one line at a time. Each sprite register can only contain the one byte of data that it needs for a single line of on-screen graphics. To draw the sprite shown above, the programmer would have to load the byte of graphics for the alien invader that corresponds with the current line on the television display and store that value in the sprite graphics register during the horizontal blank, in between the drawing of two lines. To position a sprite vertically, the programmer would have to keep track of which lines of the display have sprites on them, and compare the current line to that value in memory before drawing.

Sprite Colors

The TIA also provides a register to set sprite colors, one for each sprite (named COLUP0 and COLUP1, respectively). In early VCS games like *Combat*, sprites colors were usually set once for the entire game. In later games, programmers stored a different color value in one or both sprite color registers along with a different bitmap value. Multicolor sprites, such as the player character in Activision's *Pitfall*, allow for more visually interesting graphics. The careful observer can note color banding in most of these sprite graphics, though, which is not seen in the true bitmapped graphics of later platforms like the NES. This style of "stripe-colored" sprites is a particular trademark of VCS games.

Sprite Variations

To allow for variations in sprite graphics, the TIA offers two Number-Size registers that enforce automatic modifications to the sprites when drawn on screen (named NUSIZ0 and NUSIZ1, respectively). In particular, the programmer can change the number of sprites drawn on a single line, as well as the size of the sprites. The size of missile graphics, which are always comprised of a square shape corresponding in color to its parent sprite, are also adjustable. Adjustments to the sprites are made by setting one or more of the lowest three bits on the Number-Size register register. Table 1 shows a summary of the size and number adjustments afforded by this register.

D2	D1	D0	Description
0	0	0	one copy
0	0	1	two copies - close
0	1	0	two copies - med
0	1	1	three copies - close
1	0	0	two copies - wide
1	0	1	double size player
1	1	0	3 copies medium
1	1	1	quad sized player

Table 1: Effects of setting the VCS Number-Size registers.

The Number-Size register offers an easy way to modify the appearance and behavior of player sprites. The most transparent use of this technique is in *Combat*, which uses the Number-Size settings as the basis for many of its 27 game variations. The bi- plane and jet plane variations that double, triple, or stretch one or both sprites use the Number-Size register to accomplish what would otherwise have had to be done through complex on-the-fly graphics processing or by storing additional sprites in precious ROM — *Combat* is a 2k cartridge. For example, variation 19 is "2 vs. 2 Bi-Plane," in which each player controls two planes which fly in formation. This variation does nothing more than setting NUSIZ0 and NUSIZ1 to the binary value %00000001, which corresponds with "two copies - close" in the Number-Size register table above.

Variation 20 is "1 vs. 3 Bi-Plane," in which player one controls a large plane and player two controls three small ones in formation. This variation sets NUSIZ0 to %00100111 (quad sized player) and NUSIZ1 to %00000110 (three copies - close).

Variation 20 demonstrates the opportunities and limitations of the Number-Size registers for gameplay modification. Player 1 is at a disadvantage, since his plane is larger and therefore more vulnerable to fire. To counterbalance, this variation increases the size of the missile so that player 1 does not have to be as accurate: The third flipped bit in %00100111 increases the size of player one's missile to 4 TIA clock cycles, or roughly 4x the size of player 2's missiles. However, when player 2's sprites triple, the TIA automatically triples its missiles as well, making it even easier for player 2. A more appropriate orthogonal design approach for this variation might have been to speed up the larger player and/or his missile, thereby offsetting player 1's increased target footprint. However, to do so would have required changes in the game's logic, not just in the data settings that map variation to sprite appearance. The tradeoffs in such a decision are typical of VCS game programming.

Multiple Sprites

As we discussed above, the VCS shared the video game marketplace with coin-op arcade games, most of which were built on much more sophisticated technical infrastructures. The VCS exchanged graphical complexity and specificity of circuit design for multiple cartridge home play. But the massive popularity of arcade games like *Space Invaders* and *Pac-Man* suggested a special opportunity for the VCS: home versions of these popular coin-op games were bound to be hits.

Combat, with its tank variations based on *Tank*, showcases the hardware affordances of the VCS more clearly than almost any other game. For example, it uses two sprites, each of which fires a corresponding missile. But games like *Space Invaders* were not designed with the peculiarities of the VCS in mind. Sprites were different in many post-1977 arcade games. Notably, there were often more than two per screen!

When faced with the rows of aliens in *Space Invaders* or the fleet of ghosts that chases *Pac-Man*, VCS programmers needed a way to draw more than two sprites, even though only two one-byte registers were available.

One method comes from an exploit in the way sprites are positioned on screen horizontally. A VCS programmer positions a sprite vertically on screen by comparing a counted television line number with a variable stored in RAM to see if a sprite needs to be drawn there. This technique is grounded in the nature of the CRT television: the horizontal blank offers a natural break in screen drawing during which a few cycles of processing can be accomplished. But no similar natural mapping exists for horizontal positions on screen. To allow the VCS programmer to position sprites horizontally, the TIA exposes a set of horizontal motion registers for each of the sprites, the missiles, and the ball (named HMP0, HMP1, HMM1, and HMBL, respectively). Any object that is not intended to move must have 0 set in its corresponding register. To move an object, the programmer writes an offset value from -8 to +7 to move that object by the corresponding number of TIA color clocks. The TIA also exposes another register called HMOVE to execute changes in horizontal motion. These registers were primarily intended to be set during a vertical blank — that is, between screen draws. For example, *Combat* repositions both player and missile horizontal positions each frame, then updates variables in RAM to insure that the objects are drawn on the appropriate lines, then updates the horizontal motion registers once at the start of the frame.

Larry Kaplan, one of the first developers to work on the Stella prototype, reasoned that sprite data could be reset more frequently than once per frame.[7] Because the VCS requires the programmer to control every line of the television screen, it was also be possible to change the sprite graphics values and their horizontal positions more than once per frame. He first used this technique in *Air Sea Battle*, one of the console's launch titles. In the game, multiple rows of enemies, one per row, pass back and forth across the screen. Each player controls a turret on the ground that can be aimed and fired to destroy the enemies in the air. The multiple targets are accomplished by resetting the sprite graphics multiple times down the screen. Finally, when it is time to draw the ground, the sprite graphics and horizontal positions are reset for the player turrets.

Another variation of the horizontal movement technique helped bring *Space Invaders* to the VCS. The trademark feature of the popular arcade game was its rows of slowly descending aliens — which the TIA, of course, didn't support in any obvious way. Kaplan's *Air-Sea Battle* technique allowed multiple sprites to appear down the screen, but *Space Invaders* required multiple sprites in a horizontal line. Rick Mauer, the programmer for the VCS port of *Space Invaders*, discovered that strobing HMOVE while a line was being drawn would reposition objects immediately, even if they had already been drawn earlier in that line.[8] The TIA, lacking memory of what it has already done, will begin drawing the data from its sprite graphics registers to the screen any time that HMOVE is reset. After one row of aliens was drawn using this technique, Mauer read and wrote new sprite graphics values from ROM to create a new row of aliens with a different appearance.

These two techniques, combined with the VCS's lack of a frame buffer and subsequent requirement that the programmer draw every scanline, allowed the VCS to overcome the apparent limitation of only supporting two sprites on screen. Rather than changing both sprites and their positions every frame, one or both could be changed every line. Together, these approaches extended the game design space on the VCS, making it capable of playing games very different from the *Pong* and *Tank* arcade titles that had been the hits of the mid-1970s. The importance of these exploits was not lost on Atari executives, either. Discussing this technique in 1983, after he had become Atari VP of Product Development, Kaplan commented, "Without that single strobe, H-move, the

VCS would have died a quick death five years ago."[9]

Despite the cleverness of these techniques, both vertical positioning and horizontal strobing required sprites to move together in vertical unison. Some variations of *Air-Sea Battle* moved different enemy sprites at different rates of speed by writing new values to the horizontal motion registers, but the objects only moved horizontally, never along both horizontal and vertical axes. After the VCS port of *Space Invaders* enjoyed considerable success, partly rescuing Atari from the losses of 1977-78, the company became even more interested in arcade ports. One obvious target was *Pac-Man*, whose U.S. arcade success in 1980 made it ideal for home console adaptation. But the four *Pac-Man* monsters need move horizontally and vertically, and independent of one another, as had not been done before. Just as *Space Invaders* would have been unrecognizable without its characteristic rows of invaders, so *Pac-Man* would have been unrecognizable without its characteristic quadruplet of monsters.

To accomplish this task, Tod Frye relied on a technique called flicker. Each of the four ghosts was moved and drawn in sequence on alternating frames; Pac-Man himself is drawn every frame using the other sprite graphic register. The TIA synchronizes with an NTSC television 60 times per second, so the resulting display showed a solid Pac-Man, maze, and pellets, but ghosts that flickered on and off every quarter of a second. The phosphorescent glow of a CRT television takes a little while to fade, and the human retina retains a perceived image for a short time, so the visible effect of the flicker is slightly less pronounced than it really is. The fact that the monsters in Pac-Man were commonly referred to as "ghosts" apologized somewhat for the flicker, which suggests the dimness of an apparition. Nevertheless, the flicker technique was widely criticized by players. Later ports of games in the *Pac-Man* family, including the 1982 *Ms. Pac-Man* and the 1987 *Jr. Pac-Man*, used less visually intrusive techniques to draw the ghosts. While these last two examples have been arcade ports — providing strong, specific motivation to programmers who are charged with imitating an existing game's visual appearance and behavior — the development of original VCS games was also affected by the nature of the system's sprite capabilities and the development of techniques to exploit this capability in previously unseen ways. Not only *Air Sea Battle* but also *Adventure, Freeway, Star Wars: The Empire Strikes Back,* and many other games were developed by programmers who carefully explore and exploited the sprite drawing capabilities of the system, and who learned, directly or indirectly, from what programmers before them had done.

2006: THE NINTENDO WII

The Wii from Nintendo offers low raw processing power but an innovative controller, recalling the way that the Nintendo DS relates to the Sony PSP. We describe the Wii's controller system, which uses accelerometers and radio frequency communication with the console base to map user gestures onto a three-dimensional space. We consider how the Wii platform relates to past controller development.

Low-Power, High Activity

When Nintendo announced the Wii in 2005, calling it the Revolution at that point, it turned away from its major competitors in the console videogame industry. Competing consoles from Sony (PlayStation 3) and Microsoft (Xbox 360) focused particularly on improved graphics, including support for higher-resolution high definition (HD) TV displays. Both Sony and Microsoft's entries into this generation of machines included higher powered processors and increased RAM, providing support for larger, more detailed game worlds. Nintendo took a very different tack, not trying to keep pace with these massive improvements in processing and memory power. Whereas the Xbox 360 boasts 512MB RAM and a triple-core processor running at 3.2 GHz, the Wii's IBM processor runs at 728MHz and couples to 88MB of total RAM.[10] This is an improvement over the GameCube's 485 Mhz IBM PowerPC processor and 43MB RAM, but doesn't compare to the step up that the Xbox 360 and PlayStation 3 made from their predecessors. Instead of adding general computational power or much more highly-powered graphics, Nintendo focused on a new type of intuitive gestural interface. Nintendo also announced a focus on simpler games intended for players of all ages, suggesting that the Wii's gestural interface would afford more intuitive, facile control in games.

Influences

Physical interfaces for video games are nothing new. Understood in the broadest sense, arcade games and pinball machines had physical interfaces that required players to stand and jostle vigorously as they played. Custom physical control peripherals for the home console can be traced back to the early 1980s: Amiga's 1982 Joyboard was a plastic platform the player stood upon, rocking in all directions for control. In 1982 Atari planned an exerbike controller for the VCS codenamed "Puffer," although the device was never released commercially. LJN followed in Amiga's footsteps with the 1987 Roll 'n Rocker, a balance board controller for the Nintendo Entertainment System. Other types of physical interfaces include pad controllers and camera controllers. Pad controllers are best known today thanks to the success of *Dance Dance Revolution* but had their origins in early devices like Exus's 1987 Foot Craz for Atari VCS and Bandai/Nintendo's 1988 Power Pad for NES. The canonical camera controller is Sony's EyeToy, first introduced for PlayStation 2 in 200X, which uses computer vision to translate a player's gross motor movements into in-game game actions.

But the Wii uses different technology. Instead of converting physical movement into joystick direction as the Joyboard does, responding to floor-level touch sensors as a dance pad does, or using difference filters to detecting changing movement patterns in a video image, the Wii uses a combination of gyroscopes, accelerometers, and infrared sensors to accept user input. All of these sensors are built into the main controller, the "Wii remote." Even though Nintendo's marketing rhetoric made claims for radical innovation (as in the code name "Revolution"), both the technology and its application in games had already been tested in previous titles, mainly by Nintendo itself.

Gyroscopes and Accelerometers

Nintendo first experimented with motion controls in a title for the Game Boy Color handheld, *Kirby Tilt 'n' Tumble,* which was released in August 2000 in Japan, hitting the US market in April 2001. The well-received action/puzzle game has a gyroscope built into the cartridge and senses when the unit is tilted or jerked upward; Kirby moves left or right, or pops up into the air, in response to these movements. This technique was refined in cartridges for the GameBoy Advance (GBA). The third installation of the company's popular *WarioWare* series, *WarioWare Twisted* (Japan 2004/US 2005), offers microgames that require that the player turn, shake, and twists the entire GBA by applying yaw to the device. The second title, *Yoshi Topsy-Turvy* (Japan 2004/US 2005), adapted classic 2D platformer gameplay for a gyroscope controller. Players twist the handheld itself and also use the more traditional d-pad and button interface to move the character Yoshi through obstacles. Twisting the device alters the game world's gravitational center, allowing the player to solve physical puzzles by moving platforms or changing world's apparent floor.

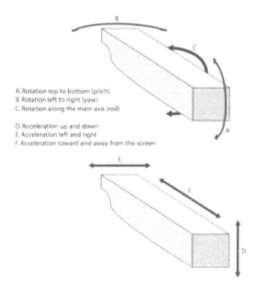

A. Rotation top to bottom (pitch)
B. Rotation left to right (yaw)
C. Rotation along the main axis (roll)

D. Acceleration up and down
E. Acceleration left and right
F. Acceleration toward and away from the screen

Figure 2: Categories of movement detected by the Wii remote.

These titles do not provide as rich a system of control as the Wii remote does, but they record how Nintendo used "cartridge hacks" as a way of prototyping gyroscopic control, both as an abstract interface principle and as a product in the market. The Wii extends the idea of earlier motion-sensitive cartridges both conceptually and technically. The Wii remote contains gyroscopes and accelerometers capable of measuring rotational movement and acceleration along the device's major axes. The basic motions shown in Figure 4 are the ones that are possible to detect and interpret in software.

In addition to the gyroscopes and accelerometers, the Wii also connects to an infrared (IR) strip sensor that detects when the Wii remote is pointed directly at the screen. The user configures the Wii for the location of the sensor — above or below the screen — and the sensor couples its readings with the top/bottom and left/right gyroscopic rotation sensors to create a pointing device. A cursor is displayed onscreen, and the Wii uses this cursor for primary selection in console menus, and some games also use it for gameplay.

Intuitive Interfaces

The Wii promises a more intuitive interface for video games. Both marketing and popular praise for the device announces that players can simply move the device as they would a prop like a tennis racket or bowling ball, and the game will respond accordingly. But the machine doesn't actually understand complex gestures that correspond directly to the basic components of real world physical activities. Instead, it understands rotation and acceleration along the axes just described. As a result, the software opportunities for the machine must either translate complex combinations and sequences of rotation and acceleration, or rely on simpler motion detection with the assumption that the player will interpret those simpler actions as more complex ones. The Wii's apparent ability to recognize gross motor gestures comes from combining combinations of the more basic motion detection methods.

The translation of the Wii's motion detection capabilities into basic gameplay mechanics are most easily observed in *WarioWare: Smooth Moves*, the first Wii game in the previously mentioned microgame-based series. In *Smooth Moves*, the player learns a series of very simple gestures, each in turn used for a set of microgames. With few exceptions, these gestures are constructed of one or two of the basic mechanical readings afforded by the device. The game gives the gestures themselves clever names (it calls them "forms," recalling basic postures in a martial art) both to telegraph interaction methods to the user and to hide the relative simplicity of the gestures.

For example, the game's first gesture, called "The Remote Control" asks the player to point the device at the screen like a remote control. This gesture mimics the recognition method of the Wii's basic pointing mode. Another gesture, called "The Umbrella," asks the user to hold the Wii remote upright and then move its tip down toward the screen at the proper time. This gesture relies on gyroscopic pitch recognition. Another gesture, called the "Pencil" asks the user to hold the Wii remote on its sides like a pencil, and to move it toward the screen. This gesture relies on toward/away from the screen acceleration recognition. Yet another gesture, called "The Mohawk," asks the user to hold the Wii remote on his head and to move his body up and down by bending at the knees. This gesture relies on up/down acceleration recognition. And another gesture, called "The Waiter," asks the user to balance the Wii remote in the palm of his hand, moving the hand on a plane parallel to the floor. This gesture relies on left/right (and sometimes also on toward/away) acceleration recognition. Much as *Combat* offers a window into the hardware affordances of the VCS, *WarioWare Smooth Moves* shows the basic capabilities of the Wii.

Despite the platform's novelty, its controller is not a magical gesture recognizer, as some critics have noted.[11] As the actual technical affordances above attest, the device is no more capable of detecting the actions a player is actually performing than one might be capable playing charades with a man behind a curtain. More "realistic" games like *Red Steel* or *The Legend of Zelda: Twilight Princess* ask the player to swing the Wii remote like a sword. But the console does not judge these gestures based on their swashbuckling quality; rather, it simply looks for up/down and left/right acceleration. Clever players might quickly realize that sword swinging can be accomplished equally well by slouching on the couch, beating the cushions occasionally with the Wii remote. Such exploits are not defects in the platform; rather, they expose the technical underpinnings of the Wii's gesture system as it is actually implemented in hardware.

Nintendo took the opportunity on two portable platforms to load cartridges with additional sensors. Finding that there were development opportunities and that players were receptive helped to justify the risks of the Wii's unusual controller scheme. While new controllers can be developed for the Wii, the irony may be that as cartridge-based R&D systems such as the Game Boy Color and Game Boy Advance disappear, there will be fewer opportunities for low-cost, per-title, *ad hoc* interface growth. The more polished Wii, with innovative interfaces built in, may not provide room for the very sort of new controller experimentation that made the platforms possible.

CONNECTING PLATFORMS, PAST AND PRESENT

The platforms discussed, and the approaches used to understand them, help to connect the technical underpinnings of new media to what has been created. The bare-bones graphics system of the VCS and the techniques developed to exploit that system in unanticipated ways has an influence on specific games and on whole genres, such as 2D platformers and shooters. The Wii shows that controllers, as well as core functions, have a history and can be considered in a platform studies approach, and that even "peripheral" elements have the potential to power innovation in game development.

The cases studies here are the result of two different platform studies approaches to two different platforms. They were not undertaken in the hopes that they would lead to a single insight about one aspect of new media. What they show, instead, is that platform studies is a rich approach that can provide a variety of insights about new media's evolution. Studies of the material history of texts have shown how the technical specifics of writing and printing technologies can inform our understanding of literary history; we believe this sort of examination is even more important in new media, which is based on complex technologies that are capable of general computation, of response to user input, of storage and retrieval of information on a large scale, and of multi-

and metamedium function. Not only do our three short studies fail to exhaust what platform studies can do in the specific cases of these platforms — they do not even begin to illustrate all the major categories of platforms or possible areas of technical focus.

EPILOGUE

We are hopeful that our efforts in platform studies will be of good use to those interested in creativity and computing, and that others in the digital media community will join us in this studies. To this end, we are happy to announce a new Platform Studies book series we are co-editing at the MIT Press. Potential writers should consult the series website at http://www.platformstudies.com.

Endnotes

1 Nick Montfort, "*Combat* in Context," *Game Studies* 6:1 (2006), http://gamestudies.org/0601/articles/montfort.

2 Tekla E. Perry and Paul Wallich, "Design case history: the Atari Video Computer System," *IEEE Spectrum* 20:3 (1983), 45-51.

3 Harold L. Vogel, *Entertainment Industry Economics* (Cambridge: Cambridge Univ. Press, 2001).

4 Martin Campell-Kelly, *From Airline Reservations to Sonic the Hedgehog: A History of the Software Industry* (Cambridge, MA: The MIT Press, 2004).

5 Scott Cohen, *Zap!: The Rise and Fall of Atari.* (New York: McGraw-Hill, 1984).

6 Perry and Wallich.

7 Ibid.

8 Ibid.

9 Ibid.

10 Marshall Brain, Jennifer Hord, "How the Wii Works." *How Stuff Works.* http://electronics.howstuffworks.com/nintendo-revolution.htm; Matt Casamassina, "IGN's Nintendo Wii FAQ," *IGN.* Sep 19. 2006, http://wii.ign.com/articles/733/733464p1.html.

11 Erik Sofge, "Nintendon't: The Case against the Wii," *Slate.* November 20, 2006, http://www.slate.com/id/2154157.

Looking Back: A Decade of Using Games to Teach History, 1996-2006

Patricia Seed

When incorporating computer games into a course on the history of European expansion in 1996, the process seemed straightforward: game reviews, historical simulations, and electronic texts. Over the course of the next decade, however, changes in computing eliminated the possibility of continuing to use the same techniques. Topics and genres of computer games that functioned well from 1996 to 1999 were virtually inoperable by 2000. To continue to use games to teach history meant confronting both significant innovations in operating systems and machines compelling a re-examination of the topics and techniques used. This presentation traces the changes and evolving tactics for one teacher for using games in the undergraduate history classroom.

During the second half of the 1990s I began to incorporate historical games as part of a course on the history of European overseas expansion. The confluence of four unrelated developments in the electronic world made this introduction possible.

At the time, U.S. computer game designers were producing a slew of historically interesting accounts of Europe's overseas expansion. Among the designers Sid Meier's *Pirates! Gold* (1993), *Colonization* (1994), *Civilization II* (1996), and Talon Soft's *Age of Sail* (1996) stood out. The 500th anniversary of Columbus voyage to the New World in 1992 may have inspired game designers to address this topic. But the game market also figured in their appearance.

During the 1990s designers and educational groups could afford to devote themselves to creating intellectually interesting historical games which could be created without enormous capital outlay, and whose profits, if any, would remain modest. Characters appeared in two dimensions, and their movements were constrained.

Secondly, to supplement these CD games, new electronic resources also became available from overseas. In 1998 several European publishers produced a number of well designed CD-ROMs containing narratives and reference material on overseas expansion. Each provided a story that students could follow as they would in a printed book. For the first time students had access to

European points of view and knowledge, formerly only available to scholars reading the original language. Oda Édition's *Navegar* (Portugal, 1998) and Uitgeverij Verloren's *The Ships of Abel Tasman* (Netherlands, 1998) stood out among these newly available resources, rich in graphic detail and resources.

Third, alongside these interesting educational games and reference sources emerged an unrelated group of software programs targeting an equally small, specialized audience. Europeans' capacity to sail around the world depended upon finding precise new data on the motions of the heavens. Astronomy has long reigned as the most popular of sciences with the general public, and n the early 1990s several useful astronomy programs appeared.

Finally, the timing was right for the students as well. By 1996, educational games had already been introduced in elementary and secondary schools, and many undergraduates then enrolled had already been exposed in school to the Carmen San Diego franchise (1985-1998) and Oregon Trail (1985, 1991).

Introducing these materials into the undergraduate classroom proved easy. I first employed the games as texts, having students "read" the game, and then used astronomy software as a virtual time machine that allowed students to see what the first South Atlantic explorers would have seen more than five hundred years ago.

In the part of the course that treated the games as texts, I assigned books and games on alternate weeks—one week a conventional reading and book review, the next week a game. For two of the "game" weeks students read the European CDs. For these electronic sources I had students write up an assessment much as they would a book review, commenting on the argument, sources, and the method of presenting historical information.

In addition, I employed other programs for historical simulation. In order to successfully sail around the world, Europeans had to develop an extensive knowledge of the position and apparent movement of the stars.

At two degrees south, fifteenth-century Portuguese maps noted that the Pole Star vanished beneath the horizon and could no longer be used to guide ships. For several previous millennia, Mediterranean sailors had used this star, and suddenly were confronted with its loss, seeing instead Portuguese scholar Pedro Nunes would later identify as "a new sky and new stars."

Starry Night served as a virtual time machine allowing students to return to the place and moment in time when this event first occurred, and Europeans were seeing unfamiliar stars in unfamiliar places and having to identify new patterns to sail by. Scholarly controversies have yet to resolve the date of the map or the exact moment when the star's fading was first noted, although it probably occurred sometime in the late 1460s. In order for students to re-experience this moment virtually, I had them set the program to latitude two degrees south a longitude off the west coast of Africa, and a day in the 1460s. Using the program, students would virtually travel back in time and space and see skies as unfamiliar to them as they had been to sailors first navigating these waters.

Like these long-ago sailors, students were forced to decide which group of celestial objects would help in navigating southward and eastward along the coast of Africa. Several would inevitably forget to consider the possibility of a nearby West African coast and, like some of the earliest explorers, would come to grief on land somewhere along the coast of Gabon. Fortunately only egos were destroyed in these accidents, not ships.

To fix the location of the stars to sail by, sailors in the 1460s were experimenting with a number of different devices, by the turn of the century settling upon what would become the first standard instrument of oceanic navigation, the nautical astrolabe, a device unfamiliar to most mariners. *The Electric Astrolabe* showed how the instrument measured the height of stars (including the sun) above the horizon. By experimenting with the virtual astrolabe students learned something of the techniques that Vasco da Gama and occasionally Christopher Columbus employed to navigate.

Students responded enthusiastically to the simulations, "book" reviews, and European CD narratives, but what worked between 1997 and 1999 ceased to function by 2001. Changes in machines and operating systems meant having to rewrite these programs. But that same shift altered the construction and marketing of games, soon rendering much of the material used in this course inoperable.

The first challenge to these games came from changes in the Microsoft operating systems. In fairly rapid succession between 1995 and 2001 Microsoft moved from Windows 95 to Windows 98, Windows Me/NT and finally XP. The European electronic reference sources became unusable on the newer operating systems, and their publishers decided not to modify the programs for further use. To this date, *Navegar* only operates on Windows 95/98, and the CD of *The Ships of Tasman* has become a print source only.

In addition to halting the use of European CD-Rom reference material many of the historical games could no longer be played on the new operating systems. *Pirates! Gold* and *Colonization* ceased functioning. Thus in short order half of the activities in the course on European expansion disappeared.

A second obstacle contributing to the cessation of these publications evolved from a major hardware innovation. Computer chips that could render three dimensional graphics on PCs appeared in 1996, and by the end of the 1990s had become inexpensive and widely accessible. Consequently designers started to craft games with three dimensional characters. Creating three dimensional characters, however, made production far more labor intensive and hence more expensive than ever before since each character or landscape required drawing multiple shapes.

To transform these European colonization games into more detailed three dimensional games would have required large infusions of capital that could not be justified for a small niche market. As a result, the small, artisanally-crafted European and American historical games ceased being produced. The only program to successfully make the transition was *Age of Sail* which was redesigned to cover a topic with national appeal, early U.S. and British naval history.

To reach the larger market, game creators shifted to expansive historical epics covering the story of human history from the Stone Age to modern times. Integrating material from many different time periods and subjects, designers correctly anticipated that they could attract large numbers of players. Fireaxis transformed *Civilization II* into a barely recognizable successor called *Civilization III*, and Microsoft likewise a converted a small *Age of Empires* into massive three dimensional production covering scores of historical epochs and with half a dozen expansion units.

The length and scale of the games had unintended consequences for their use in the classroom. While the earlier games and historical re-enactments could be carried out during a single class period, the play time of these enlarged programs meant that they would require multiple class sessions and evaluations. To see if these mass market productions could be used in the undergraduate classroom, in 1999 I began to teach a world history course employing these games.

However, the long durée approach of the *Age of Empires* and *Civilization* had many disadvantages for teaching history. In order to have the structure of the game remain constant, game designers reduced or eliminated historical specificity. The large-scale programs relied upon structural causes to develop the history of civilization: technological change, warfare, and diplomacy. And the micro-level the distinguishing quirks and differences of history, available earlier vanished under an homogenizing software engine.

With the CD ROMs disappearing and the internet not yet the searchable powerhouse it has become, alternate sources of games with greater historical accuracy. A chance meeting with a board game collector led into another world that was reversing the trend in computer gaming, by increasing the number of games with historical content. While traditionally board games developed socially competitive interactions, the medium underwent a change at the end of the nineties

In the late 1990s, several designers including Germany's legendary Reiner Kniza began to develop historical board games that explored particular moments of history sometimes in considerable detail. Historical board games proved popular and successful, garnering the top international prizes for design and sales in the tens of thousands--a trend that continues to this day.

While useful in introducing the specificity of history these games were similarly designed for a larger market, and hence tended to cover the best-known and most popular historical periods: ancient civilizations, the Roman Empire, the Renaissance, and occasionally the New World. Popular titles on ancient civilizations included *Ra* [Egypt], *Tigris and Euphrates* [Mesopotamia], the ever popular Roman Empire (*Cesar*, *Circus Maximus*)—the Renaissance (*Serenissima*— Renaissance Venice—and the self-explanatory titles *Princes of Florence*, *Medici*, and *Traders of Genoa*), and the New World (*Puerto Rico*).

However, in many of the historical board games, historical accuracy was sacrificed in the interest of making game play more exciting. *Merchants of*

Amsterdam (a game on the city's rich seventeenth-century overseas trade) contained an extremely good map of the canals of the city's commercial core, but the bidding mechanism of the game bore little resemblance to the practices of Amsterdam merchants.

In the first year teaching the combination of large-scale, multi-century computer-based historical epics and more narrowly focused board games, revealed several weaknesses. *Civilization* and *Age of Empires* provided a superficial overview of a large amount of historical information—the way in which history is taught in high schools. Both games seem far more suited to secondary schools than universities.

Another category of board game, which I did not use, also seemed appropriate for secondary education in geography. Designer Alan Moon's *10 Days in Africa* and *10 Days in Europe* replicate an "Amazing Race" competition, but do a better job of teaching actual geography. His related *Ticket to Ride* series introduces the cities of Europe and their relative locations and his *Clippers* likewise is most useful as an introduction to Pacific geography

Both the board and computer games had the additional negative consequence of once again turning students into passive observers. Neither the new historical board games nor the epic computer games provided the experience of immersion into the past such as that provided by *Starry Night* or *The Electric Astrolabe*. One of the primary goals of teachers and professors in employing games to teach has been to transform the learning experience into a more interactive one and take students into more a more active role in thinking about problems rather than memorizing outcomes.

However, overcoming the passivity of students in traditional introductory undergraduate courses seemed to present more of a challenge. I wanted students to see history as something other than a collection of canonical facts and determined outcomes—as in the repetitive --listen, note replicate— and the repeat. I wanted to allow them to experience what writing history is truly about—researching—collecting evidence—and deciding how to interpret it. I wanted to have undergraduates—most of whom had never taken a college level history course—and would probably have little future exposure to history—experience the subject the way historians do—picking facts, grouping them, sequencing, and then trying to present the facts in a single package. In that way, I hoped that students would understand how history was cobbled together and have a degree of skepticism when presented with authoritarian narratives.

If I were to have students play a design that I had created this active dimension of learning would be lost. While playing a game would no doubt be more enjoyable than listening to me lecture, it would also provide the students with a learning structure similar to that of the conventional classroom, in which they would remain passive spectators.

In order to re-engage the students as more active learners, in the next time teaching, I turned to having students design historical games based upon the events of a particular period, which they had to research, and come up with a game instead of a research paper.

In order to teach students to design games, I had to teach them how to think about contingency, likely alternative outcomes of events. In this way, the students themselves were coming up with alternative paths that history might have taken had a particular path been followed.

I found two useful means of teaching students to think about alternatives. The first employed reading historical "What If" series,[1] which provided students a way to think about alternatives such as the U.S. losing the First or Second World War (a popular game topic). Students devising military historical games were introduced to "what if" series—Students were also introduced to two other methods of thinking about historical alternatives—one through the ingenious game *Chrononauts*—which involved complicated changes which could be undone in multiple ways.

Several of the student-produced designs introduced major alternative historical trajectories by changing disease vectors in the Middle Ages or introducing rigid quarantine measures with the 1918 flu pandemic (two separate games).

Not all the games produced by students explored the alternate history possibilities, some produced art history trivia games or games designed to show why the Vikings may have begun by raiding monasteries instead of farms or towns in England and Ireland. (Vikings worshipped different gods, and encountered more valuable booty in monasteries.)

But as they created their own games, the students thought about historical alternatives and contingency, and even when they did not incorporate it in their own games; they saw how their fellow students introduced this type of thinking into their own games.

The aim of employing games in teaching world history was not to displace traditional approaches but to provide an alternative way of engage students who otherwise might not take a course in history, especially those in the sciences and engineering.

Students in engineering and the traditional sciences are accustomed to learning rule-based worlds; hence presenting the past as comprehensible within a rule-based system made the subject initially comprehensible and familiar. Scientists and engineers that I knew often indicated that they found the presentation of competing explanation and interpretations in history courses indicative of a subject that lacked both discipline and standards and which was governed not by rules but untaught, unarticulated and unexplained assumptions.

One student created a version of Chinese checkers to replicate the mercantile and commercial competition in seventeenth-century Atlantic trade. Another student took the structure of the Japanese game *Go* to reconstruct the nuclear brinksmanship of the Cold War. In other words, the most interesting projects took rules from other arenas and showed how they could explain the development of particular historical events.

However, I did discover another audience for the class, namely individuals who were interested a career in teaching, and who wanted to learn how they could create games for classrooms in which they would teach.

While the student-designed game approach has continued to be successful, some things remain lost. The rapid growth of information available on the web, alongside the rapid expansion of the first truly useful search engine (Google, founded in 1998), have been hugely positive developments. However, neither the information nor perspectives of the European CD reference sources ever transferred to the Internet. The predominant language of the Web was English, placing U.S. and British perspectives of historical events in a highly visible position and the perspectives of even other European powers receded into the background.

Having students design games does create more active and engaged students of history. But changes as yet unanticipated may bring about still further changes in the techniques of integrating games into the undergraduate classroom.

Endnotes

[1] Robert Cowley, *What ifs? of American history : eminent historians imagine what might have been* New York : G.P. Putnam's, c2003; *What if? : the world's foremost military historians imagine what might have been* New York : G.P. Putnam's Sons, c1999. A more recent book is Philip Tetlock, Richard Ned Lebow, & Geoffrey Parker, eds. *Unmaking the West : "what-if?" scenarios that rewrite world history* Ann Arbor : University of Michigan Press, 2006.

Internal Processes and Interface Effects: Three Relationships in Play

Noah Wardrip-Fruin

Eliza

In the mid-1960s Joseph Weizenbaum created a stunning piece of software. Years before HAL 9000's screen debut in *2001: A Space Odyssey*, this software, *Eliza*, made it possible to have a conversation with a computer.

Eliza ran on a computer less powerful than the one in my mobile phone, though at the time it was one of the most advanced at the Massachusetts Institute of Technology. The interface was textual. Rather than being hooked up to a monitor and a mouse — or HAL's microphone, speaker, and camera — *Eliza* was experienced through something like a typewriter, allowing the computer to type to the user and the user to type back. Yet, in the midst of all this now-primitive technology, *Eliza* was able to produce remarkable results. People could type anything they wanted to *Eliza,* and the software would respond in different ways depending on the currently-active script. Here is an example using *Eliza*'s most famous script, *Doctor,* which caused the software to parody the conversational patterns of non-directive therapist during an initial visit:

You are like my father in some ways.

WHAT RESEMBLANCE DO YOU SEE

You are not very aggressive but I think you don't want me to notice that.

WHAT MAKE YOU THINK I AM NOT VERY AGGRESSIVE

You don't argue with me.

WHY DO YOU THINK I DON'T ARGUE WITH YOU

You are afraid of me.

DOES IT PLEASE YOU TO BELIEVE I AM AFRAID OF YOU

My father is afraid of everybody.

WHAT ELSE COMES TO MIND WHEN YOU THINK OF YOUR
FATHER

Bullies.

DOES THAT HAVE ANYTHING TO DO WITH THE FACT
THAT YOUR BOYFRIEND MADE YOU COME HERE
(Weizenbaum 1966, 37)

Eliza/Doctor's final line, above, is a reference to an earlier part of the conversation in which the user wrote "my boyfriend made me come here." The resemblance between *Eliza/Doctor*'s "YOUR BOYFRIEND MADE YOU COME HERE" and the user's phrasing is not coincidental. Each script for *Eliza* is actually just a set of linguistic tricks, one of the simplest of which is to take one of the user's statements about herself and turn it back with appropriate word substitutions (in this case, "my" becomes "your" and "me" becomes "you"). Most of these tricks involve looking for key words (or small groups of words) in the user's responses, such as the "are like" in "you are like my father," above, that leads *Eliza/Doctor* to ask what resemblance the user sees.

But when we encounter the interface of a piece of software we don't necessarily get a clear picture of how it actually operates internally. And many users of *Eliza/Doctor* initially developed very mistaken ideas about its internals. They assumed that, since the interface appearance of the program could resemble something like a coherent dialogue, internally the software must be very complex. Some thought it must be something close to the fictional HAL: a computer program intelligent enough to understand and produce arbitrary human language. This happened so often, and was so striking, that computer science circles developed a specific term for this kind of misunderstanding: "the *Eliza* effect."

Play and the *Eliza* Effect

This paper is a brief look at the *Eliza* effect, and at two previously unnamed effects that can arise as we experience the interface of a digital system and build an idea of its internal operations. More specifically, this paper looks where others haven't when exploring versions of this relationship: the area of play.

Weizenbaum may have originally thought of his system as a plaything — he certainly characterized the *Doctor* script as a parody — but his attention was soon drawn to another aspect of users' interactions with *Eliza*. He came to focus on the conceptual mismatch that gives the *Eliza* effect its name, and specifically on how it could "induce powerful delusional thinking in quite normal people" (1976, 7). He wrote a book dedicated to demonstrating that the internals of computers aren't magical and that we do ourselves a disservice when we assume that human beings are so mechanical that we could, or should, have our intelligence matched by computational machines.

Weizenbaum wasn't the only one who saw the *Eliza* effect as important to address in understanding our relationship with computers. A decade after his book, Lucy Suchman published *Plans and Situated Actions* (1987), in which she sees *Eliza/Doctor* as an iconic example in human-computer interaction of what ethnomethodologist Harold Garfinkel (citing Karl Mannheim) has called the documentary method of interpretation. Specifically, that people tend to "take appearances as evidence for, or the document of, an ascribed underlying reality, while taking the reality so ascribed as a resource for the interpretation of the appearance" (23).

Yet another decade later, Janet Murray revisited *Eliza/Doctor* in *Hamlet on the Holodeck* (1997). Here she makes a crucial turn for our purposes, seeing the project in terms of media. She views *Eliza/Doctor* as a dramatic character, Weizenbaum as a kind of playwright, and the source of the *Eliza* effect in such cases as "the human propensity to suspend disbelief in the presence of a persuasive dramatic presence" (224). Here Murray makes visible what wasn't clear in previous accounts: that people enjoy *Eliza*. Initial interactions are often quite fun and engaging, and create the sense of open-ended possibilities with an interesting character.

But Murray doesn't go far into talking about ongoing interaction with *Eliza/Doctor*, just as Suchman and Weizenbaum don't. If they did, it would complicate their discussions. This is because — while the initial experience of *Eliza/Doctor* can create the surface impression of an incredibly complex internal system — sustained interaction with the system, the verbal back-and-forth, invites play ... and linguistic play with *Eliza/Doctor* quickly begins to destroy the illusion. In other words, precisely the open-ended textual interaction that helped foster the illusion of internal complexity and intelligence enables play that draws attention to the system's rote simplicity, its distance from human interaction. Before long, many exchanges begin to look like this:

You are going to repeat what I say in the form of a question

WHAT MAKES YOU THINK I AM GOING TO REPEAT WHAT YOU SAY IN THE FORM OF A QUESTION

For those of us interested in designing computationally-driven play experiences, especially those involving human characters, this presents a two-part problem. First, how can we represent something very complex — human beings — using tools that are inevitably quite simple by comparison: computational processes? Second, how can we engage audiences in playful interaction with these representations without the boom/bust of the *Eliza* effect?

Before addressing such questions, however, it is worth considering a different dilemma of digital authorship. While the *Eliza* effect produces a surface experience that can give the audience an initially-inflated view of system internals, many digital works present interfaces that fail to give audiences the opportunity to even appreciate the complexity of internal systems that are present. Rather than the boom/bust of play that the *Eliza* effect describes, these systems fail to produce an initial boom.

The *Tale-Spin* Effect

Like *Eliza*, *Tale-Spin* is a landmark of digital media. Created by James Meehan in 1976, *Tale-Spin* is the first major story generation program. It made the leap from assembling stories out of pre-defined bits (like the pages of a Choose Your Own Adventure book) to generating stories via carefully-crafted processes that operate at a fine level on story data. In *Tale-Spin*'s case, the processes simulate character reasoning and behavior, while the data defines a virtual world inhabited by the characters. As a result, while altering one page of a Choose Your Own Adventure leaves most of its story material unchanged, altering one behavior rule or fact about the world can lead to wildly different *Tale-Spin* fictions.

Tale-Spin can generate fictions with or without audience interaction. When generating with interaction, *Tale-Spin* begins by asking the audience some questions to determine the initial state of the world, especially the characters present in the story. Storytelling begins from these initially-established facts, with the audience consulted as new facts are needed to move the story forward. For example, once the characters are known and the world is established, *Tale-Spin* needs to know the identity of the main character:

THIS IS A STORY ABOUT ...
1: GEORGE BIRD 2: ARTHUR BEAR

After the audience chooses a character, *Tale-Spin* next needs to know the problem of this character that will serve as the impetus for the story:

HIS PROBLEM IS THAT HE IS ...
1: HUNGRY 2: TIRED 3: THIRSTY 4: HORNY

The opportunity for play, with *Tale-Spin*, lies in these audience choices — both in the telling of any one story and across multiple stories. Just as the audience builds up a mental model of *Eliza/Doctor* through unconstrained textual input and consideration of the software's responses, the audience of *Tale-Spin* builds one through question answering and considering both further questions and resulting stories in the context of answers given.

When the audience makes its choices, *Tale-Spin* doesn't simply record these facts about the world. In addition, internal *Tale-Spin* mechanisms draw "inferences" from the facts. For example, if it is asserted that a character is thirsty, then the inference mechanisms result in the character knowing she is thirsty, forming the goal of not being thirsty, forming a plan for reaching her goal, etc.

Some uses of inferences are relatively straightforward. It's no surprise that a thirsty character will form a plan for not being thirsty. But other uses of *Tale-Spin*'s inference mechanisms can be quite surprising. For example, *Tale-Spin* characters can use its inference mechanisms to "speculate" about the results of different courses of action. Meehan's *The Metanovel* (1976) describes a story involving such speculation, in which a hungry Arthur Bear asks George Bird to tell him the location of some honey. We learn that George believes that Arthur trusts him, and that Arthur will believe whatever he says. So George begins to use the *Tale-Spin* inference mechanisms to "imagine" other possible worlds in which Arthur believes there is honey somewhere. George draws four inferences from this, and then he follows the inferences from each of those inferences, but he doesn't find what he's after. In none of the possible worlds about which he's speculated is he any happier or less happy than he is now. Seeing no advantage in the situation for himself, he decides, based on his fundamental personality, to answer. Specifically, he decides to lie.

This is a relatively complex piece of psychological action, and certainly tells us something about George as a character. But the interface appearance of a *Tale-Spin* story never contains any information about this kind of action. For example, here is a quote provided by Meehan from a similar moment in a *Tale-Spin* story:

> Tom asked Wilma whether Wilma would tell Tom where there were some berries if Tom gave Wilma a worm. Wilma was inclined to lie to Tom. (232)

As we know from the tale of Arthur and George, a complex set of speculations and character-driven decisions took place as Wilma considered Tom's request. But all that — probably one of the most interesting parts of this story, as it is simulated inside *Tale-Spin* — is lost in the gap between the above two sentences.

No matter how creatively one plays with *Tale-Spin*, such hidden action cannot be deduced from its interface outputs. This is probably why, though *Tale-Spin* is seen as a landmark in computer science circles, it is often treated with near-ridicule in literary circles. Critics as astute as Janet Murray, Espen Aarseth, and Jay David Bolter have failed to see what makes *Tale-Spin* interesting, focusing instead on what its output looks like at the interface.

Of course, while we can call this a failure of these critics, it is probably more accurate to describe this as a failure of *Tale-Spin* itself. While *Tale-Spin*'s author created complex and interesting internal processes, he failed to make that apparent at the interface level. While playing with *Tale-Spin* actually involves setting an intricate world in motion, the audience experience is blunt and repetitive.

This situation is far from uncommon in digital media, perhaps particularly in the digital arts, where fascinating processes — drawing on inspirations ranging from John Cage to the cutting edge of computer science — are often encased in an opaque interface. In fact, this effect is at least as common as the *Eliza* effect, though I know of no term that describes it. Given this, I propose "the *Tale-Spin* effect" as a term for works that appear, at their interface, significantly less complex than they are internally.

The *Tale-Spin* effect, like the *Eliza* effect, is not only a description of audience experience — it is also a warning to authors of digital media. Just as play will unmask a simple process with more complex pretensions, so play with a fascinating system will lack all fascination if the system's operations are too-well hidden from the audience.

Luckily, the third effect I will discuss here is not a warning of this sort.

The *SimCity* Effect

In the mid-1980s, Will Wright created a landscape editor for authoring his first game, an attack helicopter simulation. Working with the editor, he had a realization: "I was having more fun making the places than I was blowing them up" (Wright 2004). From this the idea for Wright's genre-defining *SimCity* was born.

Wright realized that interacting with his terrain editor was more interesting than interacting with its outputs. In a way this is quite similar to the insight offered by the *Tale-Spin* effect: let the audience play with the most interesting parts of the system.

SimCity, of course, unlike a terrain editor, doesn't simply wait for a user to do something. Time begins passing the moment a new city is founded. A status bar tells the player what's needed next — starting with basic needs like a residential zone and a power plant and, if play succeeds for any period, ramping up to railroads, police stations, stadiums, and so on. A budding city planner can lay out spaces, but it's up to the city's virtual inhabitants to occupy them, build and rebuild, and pay the taxes that allow the city to continue to grow.

As cities grow, areas respond differently. Some may be bustling while others empty out, or never attract much interest at all. *SimCity* provides different map views that can help diagnose problems with abandoned areas. Is it too much pollution? Too much crime? Too much traffic? Players can try changing existing areas of the city (e.g., building additional roads) or create new areas with different characteristics. Observation and comparison offer insights. Why is this commercial area fully developed, while that one lies fallow? The answer is always found by trying something different and considering the results.

In other words, the process of play with *SimCity* is one of learning to understand the system's operations. Conversely, the challenge of game design is to create an interface-level experience that will make it possible for audiences to build up an appropriate model of the system internals. As Wright puts it:

> As a player, a lot of what you're trying to do is reverse engineer the simulation... The more accurately you can model that simulation in

your head, the better your strategies are going to be going forward. So what we're trying to [do] as designers is build up these mental models in the player... You've got this elaborate system with thousands of variables, and you can't just dump it on the user or else they're totally lost. (Pearce 2002)

Here, again, we lack a term for an experience. I propose "the *SimCity* effect" for this important phenomenon: a system that, through play, brings the player to an accurate understanding of the system's internal operations. Of course, the *SimCity* effect is most important to consider in cases where the system is complex, but it applies generally. *Pong* works as well as it works because it effectively communicates at the interface level its quite simple internal operations.

What is exciting about the *SimCity* effect, and about Wright's work generally, is that it helps us get at the new possibilities opened by working with computational media. *Pong* is very similar to games we play without computers, but *SimCity* is a more complex system than even the most die-hard Avalon Hill fan would want to play as a tabletop game. This ability to work with computational processes, to create complex computational systems, is the opportunity that digital media affords — and the *SimCity* effect points the way toward creating experiences of this sort that succeed for audiences.

Final thoughts

Two questions were left dangling at the end of this paper's discussion of *Eliza*. First, how can we represent human complexity using computational processes that are inevitably quite simple by comparison? Second, how can we structure play with these representations without the boom/bust of the *Eliza* effect?

A quick answer to these questions can also be drawn from the work of Will Wright, in the form of the best-selling computer game of all time: *The Sims*. This is not only the most successful game of all time, it is also a representation of human beings and their lives that successfully invites and structures play. It doesn't attempt the freeform textual dialogue of *Eliza/Doctor*, but rather has its proto-characters speak gibberish while iconic representations of conversational topics appear above their heads. In this way, and in many others, it builds on the power of the *SimCity* effect: providing the audience with a surface representation and opportunities for interaction that are at the same level of complexity as the internal system operations. It doesn't over-promise like *Eliza/Doctor*, and, unlike *Tale-Spin*, it translates the interesting complexity of its systems into audience experience.

Of course, many of us would like to play games which actually have linguistic content — in which characters actually say things in human language. Here it is useful to consider another aspect of systems: their appropriateness to what they represent. When playing an RPG such as *Oblivion*, we talk with other characters by activating them, hearing or reading lines they speak, and then choosing our own lines or topics of dialogue from a textual menu. This does a good job regarding the *SimCity* effect (the underlying system is just as simple as the surface representation) but this system is rather ill-suited to representing human characters. Each interaction with this system, each moment of play governed by it, is an abject failure compared with the smooth and compelling exploration of space provided by such games.

Given the mismatch between human complexity and most dialogue systems, how could we find a better solution? One answer, of course, is to follow Wright's lead and keep pushing forward on the complexity of the systems. But this is not a viable solution for most game designers. Perhaps better guidance could come from another designer: Jordan Mechner. His *Prince of Persia: The Sands of Time* deals elegantly with the limitations of its dialogue system by never making them available for direct interaction. Instead, the audience plays with the systems that work well when governed by the *SimCity* effect, such as acrobatic movement through the game's compelling visual spaces. Well-drawn characters exist, and speak dialogue, but their dialogue is driven by play with other systems. The main non-player character, Farah — and the player character himself — speak dialogue related to their current positions in space, progress on solving puzzles, and how these translate into forward movement in the story. The player can elicit responses from Farah by, for example, moving the player character in front of her and switching camera views to stare at her — but interaction remains firmly within the game's systems that are well-suited to representing their subjects (movement, the gaze, and combat).

Of course, such approaches will never produce the excitement we can feel during the initial moments of play under the *Eliza* effect. But such experiences can remain compelling over long periods of play, and result in characters more engaging and well-drawn because they are not founded on quick-crumbling illusions.

Meanwhile, we can continue to explore more complex models for representing human lives in the territory opened up by *The Sims*. Or, as I do in my own practice, we can explore the potential for systems that enable textual play with literary language about human relationships, rather than restricting ourselves to play with iconic graphical representations. Or, like the notable recent independent game *Façade*, we can combine more complex internal models with greater attention to the crafting of audience expectations that Janet Murray calls "scripting the interactor."

Bibliography

Aarseth, E. J. (1997). *Cybertext: Perspectives on Ergodic Literature.* Baltimore: Johns Hopkins University Press.

Bolter, J. D. (1991). *Writing Space: The Computer, Hypertext, and the History of Writing.* Mahwah, New Jersey: Lawrence Erlbaum Associates, Inc.

Murray, J. H. (1997). *Hamlet on the Holodeck.* New York: The Free Press.

Meehan, J. R. (1976). *The Metanovel: Writing Stories by Computer.* PhD thesis, Yale University.

Weizenbaum, J. (1976). *Computer Power and Human Reason: From Judgment to Calculation.* New York: W. H. Freeman.